Green Energy and Technology

More information about this series at http://www.springer.com/series/8059

Andrzej Wasiak

Modeling Energetic Efficiency of Biofuels Production

Springer

Andrzej Wasiak
Faculty of Engineering Management
Bialystok University of Technology
Białystok, Poland

ISSN 1865-3529 ISSN 1865-3537 (electronic)
Green Energy and Technology
ISBN 978-3-030-07493-7 ISBN 978-3-319-98431-5 (eBook)
https://doi.org/10.1007/978-3-319-98431-5

This Springer imprint is published by the registered company Springer Nature Switzerland AG
The registered company address is: Gewerbestrasse 11, 6330 Cham, Switzerland

Contents

Chapter 1
Introduction

During recent years, biomass became an important resource for the production of biofuels, which became an important part of the liquid fuel industry. Several advantages [1] of biofuel production and use as an energy source in various situations are usually indicated.

First of all, in many countries, it may reduce dependence upon the import of petroleum. This obviously can be achieved due to the replacement of imported fossil fuel by the biofuel derived from domestic biological resources. Moreover, a variety of those resources are available, starting from edible parts of plants through nonedible ones or just wastes from, e.g., food industry.

The second advantage can be associated with the possibility of leverage of limited supply of fossil resources. In spite of diverse opinions about the amounts of resources still available or expected to be developed in the future, it seems undisputable that the availability of all resources existing on the Earth is limited. Also, an increase in the world's demand for primary energy is expected [2]. Therefore, the extension of energy resources beyond the available fossil fuels is valuable.

The use of fossil fuels is undoubtedly connected with the emission of carbon dioxide, which is frequently considered as a cause for global warming [e.g., 3, 4]. The technological as well as economic difficulties connected to CO_2 mitigation brings some authors [5] to the conclusion of the need for limitation of future average world's energy consumption to the level of approximately 63 GJ/capita/year (what corresponds to the available power of 2 kW/capita). This limit is much lower than the actual primary energy use of some countries, and lower than the actual world's average [2].

Since biofuels are considered as, at least partial, the replacement of fossil fuels, and they are considered as substances that emit this amount of carbon dioxide which was earlier absorbed by the plants during their life. It is also supposed that future plants will also absorb that the amount of CO_2 during their growth. Consequently, at least in equilibrium, those acts of emission and absorption do not bring additional carbon dioxide from the Earth deposits.

A. Wasiak, *Modeling Energetic Efficiency of Biofuels Production*, Green Energy and Technology, https://doi.org/10.1007/978-3-319-98431-5_1

Several technologies enabling the conversion of various types of biomass onto biofuels have already been elaborated. Those technologies involve several biomass resources, various forms of biofuels and a number of technologies converting the specific form of biomass into biofuel. A typical situation in all technologies of conversion is the need for some amounts of energy that has to be provided in order to facilitate conversion processes.

The typical structure of a biofuel production system contains the coupling between agricultural and industrial subsystems by flows of substances undergoing conversion on subsequent steps, and fluxes of energy provided to the steps, on which the conversion occurs. Since the agricultural production is distributed on the relatively large area as compared to industrial production the needs for transportation of machines and materials across the fields, as well as the transportation of the crops between the plantations and industrial processing facility occurs. This transport, in addition to conversion processes, also consumes some energy, increasing the total energy consumption within the production system.

In all cases, however, it is important that the processes occurring in the whole production system should not consume more energy than the amount of energy that can be obtained in the form of final biofuel. Otherwise, no gain of energy is achieved. The main aim of the present book is to present the investigations on energy balance in biofuel production systems, and the effects caused by individual processes occurring in various parts of production systems.

References

1. Sheehan, J., Camobreco V., Duffield J., Graboski M., Shapouri H.: Life cycle inventory of biodiesel and petroleum diesel for use in an urban bus. NREL/SR-580-24089 UC Category 1503
2. Moriarty, P., Honnery, D.: What is the global potential for renewable energy? Renew. Sustain. Energy Rev. **16**, 244–252 (2012)
3. Revelle, R.: Carbon dioxide and other greenhouse gases in ocean, atmosphere, and biosphere, and future climatic impacts. Stud. Environ. Sci. **26**, 405–4284 (1986)
4. Beck E. G.: 180 years of atmospheric CO_2 gas analysis by chemical methods, Energy Environ. **18**(2) (2007)
5. Spreng D.: Distribution of energy consumption and the 2000 W/capita target. Energy Policy. **33**, 1905–1911 (2005)

Chapter 2
Biomass Resources for Biofuel Production

2.1 Types of Biomass

Depending on its origin, biomass may exhibit very different properties, and, therefore, require quite different treatments in conversion into useful products. In addition, the sources of biomass are very different. Some sources of biomass are based on the wild life of plants, and other organisms. This can be exemplified by wild forests or aquatic organisms, especially those living in oceans. The other type is cultivated. In such a case, the stand for useful crops is usually prepared, the organisms carefully protected, and finally collected. Usually, except for the useful part, some part of the biomass is less suitable and consists of some wastes, which require other type of utilization, different from that of the main crop. Figure 2.1 gives an example of several types of biomass resources that are used for fuel production. The figure contains only operations that are needed to obtain biomass prior to conversion of biofuels. The energetic crops are various plants that are cultivated especially to produce some kind of fuel. The majority of such plants are land species; however, in recent times, some aquatic species are being under consideration, at least at the experimental level. The land species are mostly s.c. short rotation plants, which means they are produced in one to few years cycles.

Such plantations require various forms of tillage, application of fertilizers, crop protection means, and obviously harvesting.

The other types of biomass that can be used for fuel production are agricultural residues or wastes—the amounts of biomass remaining after the removal of edible parts of the crops (fruits, grains, etc.). Utilization of those parts of biomass requires the choice of processes well suited to the specific characteristics of particular by-products that have to be converted to an energy carrier. It should be also mentioned that quite frequently edible parts of the plants are also used for biofuel production, e.g., rapeseed or maize grains.

A. Wasiak, *Modeling Energetic Efficiency of Biofuels Production*, Green Energy and Technology, https://doi.org/10.1007/978-3-319-98431-5_2

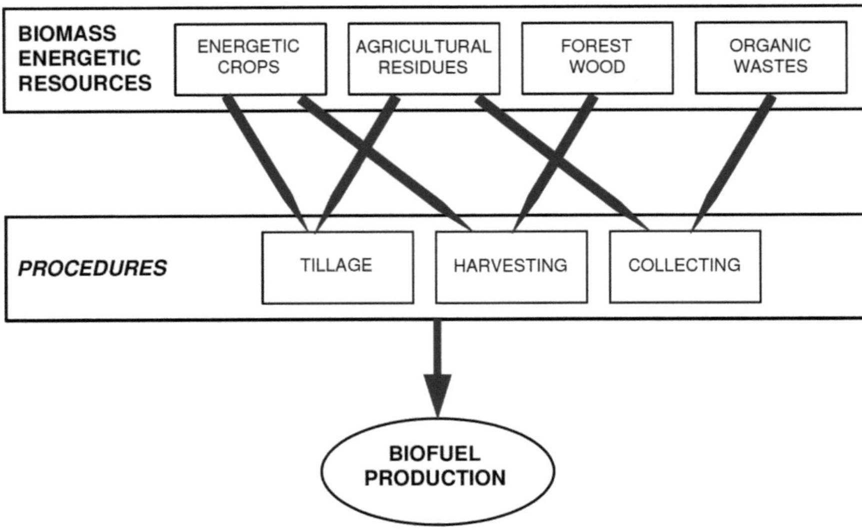

Fig. 2.1 Examples of biomass types and operations preceding biofuel production

2.2 Forests as Wood Resource

Wood, being exploited already in ancient times [1] to a large extent is used as construction material, either after very simple treatment as timber or as a raw material for the production of more advanced wood-derived materials like plywood, several types of panels, veneer, etc. A substantial amount of harvested wood is consumed for paper production. The lower quality wood fractions are used as an energy source. This wood is either used, after simple treatment—chopping, as a firewood or undergoes more advanced processing onto the other types of wood-derived fuels.

Chemical composition [2] of wood may slightly differ between various species, but in general, the composition corresponds to the data listed in Table 2.1.

Wood is generally distinguished as softwood, obtained from coniferous trees, and hardwood—from broad-leaved trees. Both types differ in structure and properties. Due to its form, the fuelwood [3] is measured for trade in several volume units:

Table 2.1 Typical composition of wood

Table	Component	Mass %
1	Carbon	51
2	Hydrogen	6
3	Oxygen	42
4	Ash	0.5
5	Water	10–65

Table 2.2 Basic density of several air-dried wood species

Tree species	Basic density of oven dry wood (kg/m^3)
Silver fir	410
Poplar	410
Norway spruce	430
Scots pine	490
Larch	550
Birch	610
Oak	650
Beech	680
Black locust	730

Source N. Krajnc; Wood Fuel Handbook; Food and Agriculture Organization of United Nations, Pristina 2015

- Solid volume—which is the volume of individual pieces, is measured in solid cubic meter (scm—m^3).
- Stacked volume—which corresponds to the one cubic meter of neatly stacked logs of wood including air gaps between logs, is measured in stacked cubic meter (stcm—stacked m^3).
- Bulk volume—which is applied to wood chips, sawdust, or wood pieces attaining one cubic meter including air gaps (bcm—bulk m^3).

The above-described notions are related to the definitions of density. The most important are: the particle density, which is the real density of the material [kg/m3], and bulk density, which depending on the situation, is referred either to stacked volume [kg/scm] or to bulk volume [kg/bcm]. In the forests economy also, the term "Basic density" is used which denotes the ratio of the mass of wood measured on the dry basis to the solid volume measured on sc. green basis. Basic densities of several species are listed in Table 2.2.

Values of bulk density play an important role in the determination of truck capacity for wood transportation.

The density of "green" wood (containing some fraction of water) is higher than that of a dried piece of the same species [4]. In forestry and timber industry, several definitions of density coexist. The green density, d_g, is defined as the ratio green mass, m_g, to the green volume, v_g, of wood, i.e.,

$$d_g = \frac{m_g}{v_g} \quad (2.1)$$

while the oven dry density, d_o, is the ratio of the mass of the oven dry wood, m_o, to the same green volume:

$$d_o = \frac{m_o}{v_g} \quad (2.2)$$

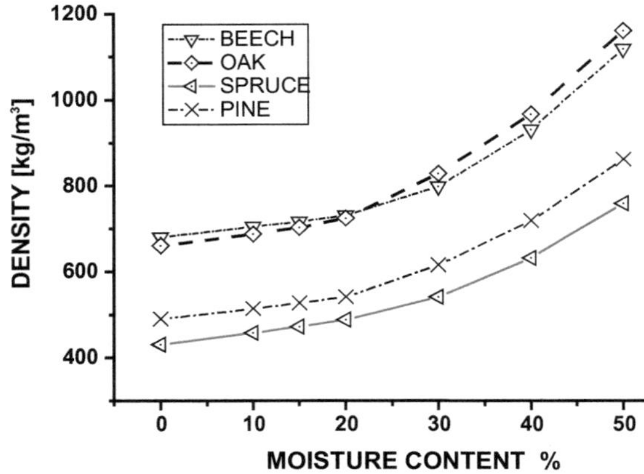

Fig. 2.2 Round wood stacked density for several species of wood as dependent upon moisture content

As indicated in Table 2.1, wood might contain a substantial amount of water. At least two measures of wood humidity are frequently used: first, thewater content, c_w, is defined as

$$c_w = \frac{m_w - m_o}{m_w} \times 100\% = \frac{m_{\text{water}}}{m_{\text{wet wood}}} \times 100\% \tag{2.3}$$

and wood humidity

$$u = \frac{m_w - m_o}{m_o} \, u = \frac{m_w - m_o}{m_o} \times 100\% = \frac{m_{\text{contained water}}}{m_{\text{drywood}}} \times 100\% \tag{2.4}$$

where m_w—is the mass of wet wood, m_o—is the mass of oven dry wood.

The humidity of wood affects substantially its properties, e.g., the density. The work [3] reports several data concerning the effect of moisture content on the stacked density of several species of wood. Figure 2.2 (plotted on the basis of the data from the work mentioned above) presents the stacked density of round logs, while Fig. 2.3 presents a similar plot for pinewood samples of various shapes, i.e., round wood, chopped logwoods, and wood chips.

It is seen that the density of dry woods as well as humid ones is different for individual species, the dependence upon moisture content shows similar for all species' nonlinear shape. The nonlinear behavior is probably due to nonuniform distribution of moisture in the volume of wood.

As seen in Fig. 2.3, various levels of dispersion strongly affect the stacked density of wood samples. Obviously, wood chips show the lowest density, and less pronounced nonlinearity of moisture dependence.

According to [4], the density of "green", the newly felled wood shows quite high humidity, and in a complex manner is dependent upon the height of the tree. As an example, Fig. 2.4 shows the dependence of density on the relative height of the tree in "green" state as compared to the oven-dried state.

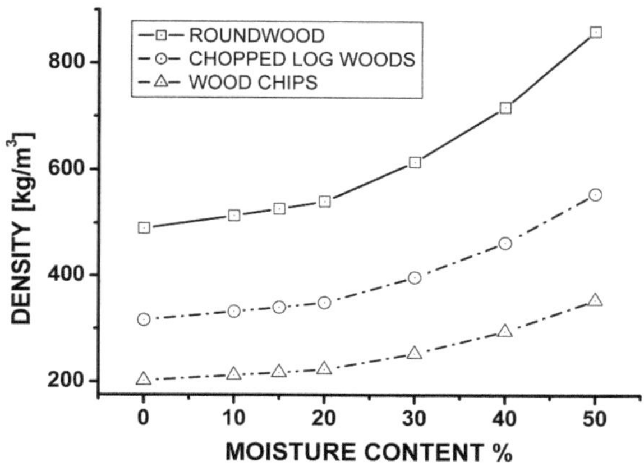

Fig. 2.3 The effect of moisture content on the stacked density of various forms of pine wood samples

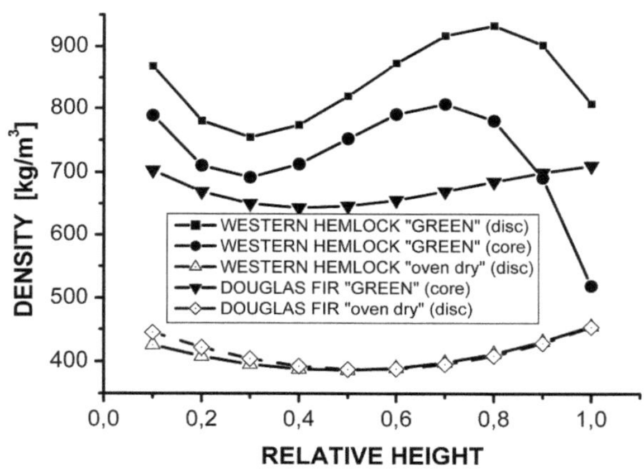

Fig. 2.4 Density of "green" and "oven dry" samples of two species of wood. *Source* Calculated on the basis of data from [4]

Table 2.3 Average density of "green" and "oven dry" samples of two species of wood

Species	Humidity	Density [kg/m^3]
Western hemlock	Green	724.9–843.5
	Oven dry	408.9
Douglas fir	Green	673.5
	Oven dry	412.5

Source Calculated based on data from reference [4]

It is seen that the "green" wood shows quite large differences in density along the height of the tree. The authors of paper [4] do not report numerical values of humidity, but show the dependence illustrating that variations of density are proportional to the local humidity of wood logs. The average density calculated from the dependencies shown in Fig. 2.2 are listed in Table 2.3 showing indicating that "green" wood shows the density even twice as large as the dry material.

Table 2.4 Example of energy consumption by ground-based harvesting systems [MJ/m^3]

	Felling	Skidding	Processing	Total
Full-tree (Clear-cutting)	28.84	23.84	26.92	79.98
Cut-to-length (Clear-cutting)	48.07	25.00		73.06
Cut-to-length (Thinning)	48.07	26.92		78.06
Full-tree (Thinning)	68.83	30.00	26.92	126.12

Again, this factor is important with respect to the transport of harvested wood to the places where it will be converted into useful products.

Forests are the main sources of wood harvested for all types of use. Actually only about 20% of global forests are the natural ones [5], while the others are plantation forests, usually resulting from reforestation and being subjected to various operations of silviculture management. Those operations are different in the cases of various climatic zones, types of forests, etc.

Concerning energy consumption needed for forest maintenance (and harvesting), it is difficult to achieve good estimation because of the very long period of forest production. During this (normally lasting for 30–50 years) period of time, many technological changes have occured due to which recovery of data forming the "energetic history" of a particular area of forest is almost impossible. The mostly credible data, actually available, concern the procedures related to harvesting. According to the work [6], several methods of harvesting are used in silvicultural procedures. The tree-length method consists of tree felling, delimbing followed by topping to the required log lengths assuring standard diameters of the small end. The whole tree method, in turn, involves tree felling followed by an extraction to the roadside of the whole tree with all branches not being separated from the stem. Subsequent method called cut-to-length method involves operations like felling, delimbing, and crosscutting to log lengths between 3–8 m and 4–6 m. Finally, the chipping method consists in grinding or chipping directly at the felling site followed by transport of ground material outside of the site.

The fuel consumption in the operations indicated above was estimated in the work [7]. The corresponding energy consumption calculated by the present author is shown in Table 2.4. It is seen that three first methods listed in the table show quite similar values of energy consumption, while the last, i.e., full-tree method applied to the thinning operations requires substantially more energy.

Since the types of machines being used, and their individual fuel consumption are not specified, it is difficult to uncover the reasons for the difference observed.

Another report [8] shows a more detailed description of processes including fuel consumption, what gave to the present author a possibility to estimate energy

Table 2.5 Energy consumption in various harvesting systems

Harvesting system	Machines used	MJ/m^3
Cut-to-length (Thinning)	FB and GSK	88.43
Cut-to-length (Clear-cutting)	FB and GSK	69.21
Clear-cutting (Log-length)	FB, GSK, LD and PRC	82.66
Clear-cutting—Thinning (Log-length)	FB, GSK, LD and PRC	105.73
Clear-cutting (Tree-length)	FB, GSK, and LD	69.21
Thinning (Tree-length)	FB, GSK, and LD	73.05

**FB* (Feller-buncher), *GSK* (Grapple skidder), *LD* (Loader), *PRC* (Processor)

Table 2.6 Fuel consumption and corresponding energy consumption for the case of Poland

Operation	Energy consumption MJ/m^3
Harvesting	30.76
Forwarding (total)	21.91
Loading	11.53
Driving	6.15
Unloading	4.22
Road transport	28.06
Total	102.66

consumption for the sets of machines used for specified methods of harvesting. The results of computations are given in Table 2.5.

Results shown in Table 2.5 are slightly higher than those placed in the previous table. Again, it is seen that thinning operations consume rather more energy than the others. It is worth to compare the above data with the other publication [9] that report the data determined in forests in Poland. The reported values of fuel consumption have been used by the present author to compute energy consumption using the calorific value of a diesel fuel. The data reported in Table 2.6 show surprisingly low values of the amount of energy needed for the harvesting of one cubic meter of wood. The operations of felling, delimbing, and further processing of the delimbed stems were performed with John Deere 1270E harvester, while logs were extracted and the stack was formed by means of John Deere 1210E forwarder. Finally, for transportation of timber from the forest to the lumberyard, the vehicle MAN TGA was used.

The authors have discussed differences between their results, and the other existing in the literature, claiming that it may be a result of the complexity of local conditions affecting each practical situation.

An important characteristic for a forest as a source of wood is the wood content in a specified volume, s.c. stocking of the specified forest area, usually measured in m^3/ha. This value is very much dependent on the age of the forest, as well as many other factors like species growing in particular forest, environmental conditions, etc.

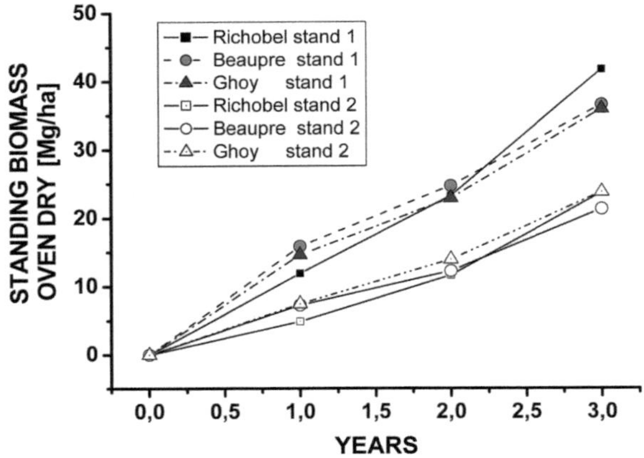

Fig. 2.5 Biomass yield from three varieties of poplar grown on two sites

Values of stocking of the forest of productive age vary between 300 m^3/ha reported in Poland [10], and 683 m^3/ha for 90 years old forest reported in England [11]. Obviously, the annual allowable cut is much smaller and depends on many factors of forest management.

2.3 Short Rotation Crops

The notion of short rotation crops is usually associated with manmade plantations of some species [12]. Several cases may be distinguished: short rotation forests, short rotation coppices, and short rotation perennial grasses. Plantations are already popular in a wide variety of countries over the Globe [13, 14]. Depending on the climatic zone, different plants are cultivated. Each type of plantation requires individual treatment, and if the rotation period is correctly chosen, provides a specific yield of biomass. In the short rotation forestry, the rotation periods depend on future application of produced wood. Rotation length of 5–10 years usually serves in energetic biomass production, rotation length between 10 and 20 years produces wood for pulpwood production, whereas 20–40 years rotation is required for timber [15]. Yields of individual plantations of short rotation coppices may depend on the location of plantation as well as the choice of particular cultivar. Examples of such dependencies are given in Figs. 2.5 and 2.6. Both figures are drawn based on the data published in [16] for plantations located in England.

Figure 2.5 shows the dependence of the yield of oven-dried biomass (in Mg/ha) as the function of the time of plantation growth for three varieties of poplar. The plot demonstrates that the rate of growth is more affected by the choice of plantation site than by the choice of the plant variety. Figure 2.6 shows a similar dependence for willow varieties. In this case, the individual dependencies show some level of irregularity, which can be associated rather with different behaviors of plant vari-

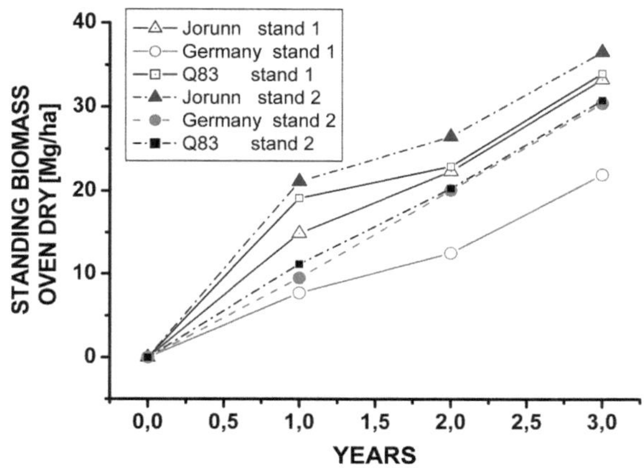

Fig. 2.6 Biomass yield from three varieties of willow grown on two sites

eties, and maybe external conditions, than the difference between plantation sites. Presented data seem to indicate that the highest increment in biomass growth is achieved between second and third production year.

Different behaviors are demonstrated in [17]. This work reports annual yields of the oven-dried biomass equal to 15.1 Mg/ha/a in the first year, followed by an increase to 22 Mg/ha/a in the second year, and later a decrease in following years: 17.2 Mg/ha/a in the third, 16.9—in fourth, and 14 Mg/ha/a in the fifth year.

Interesting results concerning black locust growth in a plantation located in Hungary are given in [18].

Several varieties of black locust were examined showing relatively close results. The chosen example is one of the more efficient ones. The curve presented in Fig. 2.7 is calculated on the basis of data presented in the original paper. It is seen that rather slow growth occurs during the first 3 years. The next period, between third and seventh year, corresponds to the highest rate of growth, (it can be roughly approximated by a linear relationship with an increment about 14.8 Mg/ha/a), followed by a decrease that cannot be quantitatively estimated due to the lack of measured data.

Unfortunately, not so many data are published with respect to energy consumption during planting and harvesting of short rotation crops. As far as short rotation forestry is concerned, one might expect some energy consumption during planting and maintenance, and during harvesting. The latter is probably similar to the energy consumption observed in harvesting of natural forests. An estimation of contributions of initial operations appears rather difficult, especially for forests with higher rotation length because many technological changes occur during the growth of a particular piece of forest. For the case of short rotation coppices, not so drastic technological changes might occur during subsequent production cycles. Some conclusions can be drawn from life cycle analysis (LCA) computations, performed for an eucalyptus plantation with 10 years cutting cycle [19]. The analysis, including production of cutting in biotechnology laboratory, site preparation, weed control, fertilization, planting, stump removal at the end of plantation life, harvesting (con-

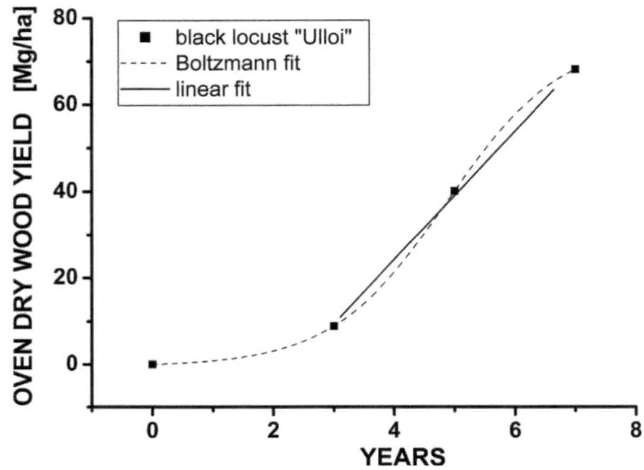

Fig. 2.7 Oven dry biomass yield from the plantation of black locust

tain felling, hauling, chipping, and machine transportation), followed by wood chips transportation, gave the ratio of energy delivered to the nonrenewable energy consumed being of the order of 11–13. The paper [20], gives, in turn, an estimation that for short rotation poplar plantation the ratio of energy consumed to energy delivered varies from 1/16 to 1/26. The authors of the work [21] on the basis of Gross Energy Requirement methodology estimated for 2 years cycle short rotation coppice energy consumption of 456 and 301 MJ/$t_{dm,}$ for 5 years cycle of medium rotation coppice. The same authors estimate energy consumption for wood chips transportation equal to 42.1 MJ/$t_{dm,}$ for the 70 km distance, and 601.9 MJ/$t_{dm,}$ for 450 km distance.

Energy consumption in willow plantations has been studied in several papers. As an example, some of the data extracted from the works [22] and [23] are presented in Table 2.7. The data from [22] the first two columns in the Table, are given for plantations located in Ireland and UK, while the data presented in columns 4 and 5 concern two plantations located in Poland. The big differences between the plantations are clearly visible. The differences result of the choice of operations in the cultivation process, as well as of consideration of other contributions in the energy consumption estimation (e.g. field transport and plantation liquidation).

Plantation working cycle in [22] is considered 23 years, and contains 7 harvesting cycles after the coppicing that occur at the end of the first season. The density of plants is not reported. In the work [23], in turn, 19 years of cultivation is predicted.

For the case of the plantation I, the density of plants is 48,000 of plants per hectare, and annual harvesting cycle, while in the case of plantation II the density is 25,000 of plants per hectare, and triennial harvesting cycle. Moreover, in the energy balance, the contribution of man's manual labor is considered. This contribution is quite large. According to the value published in [24], given as a reference in [23] the energetic equivalent of one man-hour is 100 MJ. This value is almost three times higher than the estimation given in [25]. Both papers [22] and [23] report indirect energy consumption connected with the use of fertilizers, crop protection means, etc.

Table 2.7 Energy consumption in three cases of willow plantation establishment

Operation	Total energy input averaged per annum MJ/ha/a	Operation	Total energy input averaged per annum plantation I MJ/ha/a	Total energy input averaged per annum plantation II MJ/ha/a
Plowing	40	Winter ploughing	76.8	76.8
		Disking (2×)	61	61
Harrowing	7	Harrowing (2×)	57.5	57.5
		Marking of planting lines	30.6	30.6
Planting	31	Cuttings and manual planting	709.9	367.6
Coppicing	13	Plants harvest after the first year	51.4	51.4
Fertilizer application	26			
Lime application	5			
Fertilizer application	26			
		Herbicide and spraying (soil herbicide)	45	45
		Weeding (2×)	61.6	61.6
Herbicide application	28	Spraying (glyphosate)	37.1	37.1
Insecticide application	4			
		Field transport	34.8	34.8
		Plantation liquidation	647.6	411.9
Total	180	Total	1813.3	1235.3

Columns 1 and 2 data based on the work [22], columns 3–5 data based on the work [23]; details are given in the text

In the paper [22], the use of 350 kW Class Jaguar Salix forage harvester is used for harvesting, which requires the energy consumption of 2520 MJ/ha. The energy consumption for drying willow chips estimated on the basis of latent heat of water evaporation, and humidity of the chips is estimated as equal to 18.8 GJ/ha. Finally, the total energy input (direct + indirect) for willow chips production is reported as 9257 MJ/ha/a, while energy that can be obtained from willow chips is estimated as 187–224 GJ/ha/a.

The paper [23] for the case of plantation I reports the use of JF Double Z20 SILO COMBINE coupled to a tractor, and for plantation II—the Class Jaguar 830. The quite large difference between minimum and maximum crop yield is also reported. Finally, the total energy consumption for plantation I is reported as 34.4 GJ/ha, whereas for plantation II—23.5 GJ/ha The energy content is 264.4 GJ/ha/a for the maximum, and 47.5 GJ/ha/a for the minimum crop yield (average 203.1 GJ/ha/a) for plantation I, and 461.4 GJ/ha/a for the maximum, and 19.1 GJ/ha/a for the minimum crop yield from plantation II.

The lack of detailed description of calculation procedures makes practically impossible to discover the sources of the differences observed. Important factors like details of the technology, climatic differences, local differences in indirect energy consumption, as well as different computation procedures may be responsible for different results. Anyhow, those results suggest the needs for further, systematic research using possibly standardized procedures.

An attempt at similar estimations was also recently made in [26]. Since a small error occurred in the original paper, the corrected and extended results are presented here. The purely theoretical considerations are made based on the assumed tools used, and their technical parameters applied to the estimation of their performance in the field. Two sizes of fields, namely 1 and 100 ha, were assumed, as well as the use of different sets of machines was taken into account. The original paper was confined to the plantation establishment until the first cutting (coppicing). The extension given here takes into account 20 years of existence with 10 biannual harvesting cycles starting after 2 years of initial period. The amounts of energy consumed during operations were computed according to the following formula:

$$E_{\mathrm{i}} = [A/\mathrm{Eff}_{\mathrm{i}}]\omega_{\mathrm{i}} W \tag{2.5}$$

where E_i—is energy [J] consumed in an individual operation, A—is the field area [ha], ω_{i}—is fuel consumption by the ith machine, $\mathrm{Eff}_{\mathrm{i}}$—productivity of the ith machine [ha/h], and W—calorific value of the fuel being used [J/dm^3].

Obviously, the amount of energy consumed in operations being repeated n times during life cycle of plantation can be expressed as

$$E_{\mathrm{i,n}} = nE_{\mathrm{i}} \tag{2.6}$$

Finally, the average energy consumption for each operation will assume the following form:

$$E_{\mathrm{av,i}} = n[A/\mathrm{Eff}_{\mathrm{i}}][\omega_{\mathrm{i}} W/\mathrm{AN}] \tag{2.7}$$

where N—is the lifetime of plantation [years].

Since biannual harvesting is taken into account, it might be useful to use biannual average which amounts to the doubled value of the above, i.e., $\mathrm{E}_{\mathrm{2av,i}} = 2_{\mathrm{Eav,i.}}$

The corresponding contributions to the total consumed energy are presented in Table 2.8 for the 1 ha field, and in Table 2.9 for 100 ha field. Names and characteristics of the machines used are given in both tables correspondingly. Obviously, the values of total energy consumption are sums over all operations. It should also be mentioned

Table 2.8 Energy consumption on 1 ha field

Operation	Machine characteristics	Number of operations	Fuel consumption in one operation [dm^3]	Total energy consumption in *n* operations [GJ]	The energy consumption for one operation [GJ]	Average energy consumption per annum [GJ/ha/a]
Planting	Manual	1	–	–	–	–
Plowing	Plough Unia Grudziądz 100 B $\omega = 12$ dm^3/h Eff = 0.6 ha/h	1	20	0.7	0.7	0.035
Cultivation	Cultivator Unia Grudzi0105dz ARESL/S $\omega = 11$ dm^3/h Eff = 1.6 ha/h	1	6.87	0.3	0.3	0.015
Spraying	Sprayer Pilmet 300LM $\omega = 8$ dm^3/h Eff = 4 ha/h	10	2	0.1	0.1	0.005
Fertilization	Fertilizer spreader SIPMA RN 410 ANTEK $\omega = 8$ dm^3/h Eff = 3 ha/h	10	2.67	1	0.1	0.05
Cutting	Mower Husqvarna 555FXT $\omega = 1.36$ dm^3/h Eff = 0024 ha/h	10	56.7	20	2	1
Transport	Trailer PRONAR with tractor	10	18	6,3	0.63	0.315
			Total	28.4	3.83	1.42

that in the case of small (1 ha) field, the manual operation during planting is assumed, and no energy contribution is taken into account. This approach is connected with the idea that the contributions of energy consumption by technical means are more important for estimation of fossil fuel replacement with biofuels. The values seen in the tables seem to be smaller than in the papers earlier discussed. This might be the result of "idealization" of machine's performances. In addition, the value of averaged energy consumption is higher for small plantation than that for the large one. This may be the result of the choice of too big machines (in spite that they are chosen rather small) for the 1 ha field, which causes not optimal utilization of the power of that machines. This result can be considered as kind of a warning indicating the optimization of the machines power as an important factor of energetic economy.

2.4 Agriculture

Wide varieties of annual plants are worldwide cultivated in order to convert to energy at least some part of the plant. The kind of a plant being cultivated mainly depends on climatic conditions in that particular area. The analysis of global production, given in [27], indicates that yield of a plantation, and consequently the amount of energy that can be obtained can be very different. The list of plants commonly used for the production of two typical biofuels (i.e., biodiesel and bioethanol) as well as the mean values of the amounts of energy that can be obtained from the metric ton of the crop of particular plant are given in Table 2.10, based on the data from [27]. In the case

Table 2.9 Energy consumption on 100 ha field

Operation	Machine characteristics	Number of operations	Fuel consumption in one operation [dm^3]	Total in all operations [GJ]	Energy consumption for one operation [GJ]	Average per annum [GJ/ha/a]
Planting	Planting machine Spapperi TP $\omega = 12$ dm^3/h Eff = 3 ha/h	1	3333.3	116.7	116.7	0.058
Plowing	Plough $\omega = 22.5$ dm^3/h Eff = 3 ha/h	1	1250	43.8	43.8	0.022
Cultivation	Cultivator Agro Masz 5.6 m $\omega = 21$ dm^3/h Eff = 3 ha/h	1	875	30.7	30.7	0.015
Spraying	Sprayer Pilmet EuropaXL 3000 l $\omega = 12$ dm^3/h Eff = 3 ha/h	10	120	42	4.2	0.021
Fertilization	Fertilizer spreader AMAZONE ZA-M ultra $\omega = 14$ dm^3/h Eff = 3 ha/h	10	100	35	3.5	0.018
Cutting	Forage harvester Claas 940 $\omega = 55$ dm^3/h Eff = 3 ha/h	10	2291.6	803	80.3	0.402
Transport	Trailer PRONAR with tractor	10	600	210	21	0.11
			Total	1281.2	300,2	0.641

of biodiesel, the amount of energy derivable from 1 Mg of the crop varies between 3.6 and 15.3 GJ/Mg, while in the case of bioethanol those limits are between 1.8 and 9.1. Obviously, not all plants can be cultivated in any climatic zone, and in any particular soil, so not always there are many choices. The yield of a plantation may also depend on various factors leading to a distribution of yields that differentiate various production sites in a given region. The average values of amounts of energy obtained from 1 hectare of plantation calculated on the basis of the data cited in [27] are presented in Table 2.11.

As seen in Table 2.11, the energy yield from the unit of plantation area is quite different, and usually higher in the case of bioethanol than that of biodiesel. In both cases, however, the correct choice of a plant may be an important issue. It has to be pointed out that some authors [28] indicate that there still are reserves in the possible plantations yields, which can be utilized with an intensification of agricultural technology. Some theoretical calculations [29] also indicate that the biophysical limits of plant yield are not yet achieved, and consequently some improvements of the plantation's yield may be expected.

Table 2.10 Typical plants used for biodiesel and bioethanol production, and the amount of energy that can be obtained from 1 Mg (metric ton) of the crop

Crop	Energy in form of biodiesel	Crop	Energy in form of ethanol
	GJ/Mg		GJ/Mg
Castor	13.6	Barley	5.2
Coconut	4.5	Cassava	3.8
Cotton	3.6	Maize	8.7
Groundnut	10.7	Potato	2.4
Mustard	12.8	Rice	9.1
Oil palm	7.8	Sorghum	8.5
Rapeseed	13.6	Sugarbeet	2.2
Sesame	15.3	Sugarcane	1.8
Soybean	6.4	Sweet Potato	2.7
Sunflower	14.5	Wheat	8.3

Table 2.11 Energy yields for two types of biofuels, and various plants

Fuel	Plant	GJ/ha
Bioethanol	Sugarcane	126.6–130.7
	Sugarbeet	105.5–141
	Cassava	~1
	Maize	65.5–69.9
	Wheat	52.8–54.7
	Barley	~23
	Sweet sorghum	~74
	Oil palm	155.7–164.4
Biodiesel	Coconut	~74
	Rapeseed	33–41.6
	Sunflower	26.6–34.6
	Soybean	18.2–20.8
	Groundnut	~29
	Castor	~28

Variants of agricultural production technology may lead to the differences in energy consumed by a specific plant. A comparison of amounts of energy consumed during production of a biofuel with the amount of energy obtained in the form of that biofuel brings substantial differences in estimation of contributions from various production steps. For example, the work [30] estimates the energy consumption for production of ethanol from cassava in agricultural and industrial operations being 1.54 and 11.76 MJ/l correspondingly, what together gives 13.3 MJ/l, while the calorific value of ethanol amounts to 23.37 MJ/l. Consequently, the ratio of the energy consumed to the energy produced amounts to 1.76 for the whole process.

The detailed analysis of energy consumption during subsequent steps of maize cultivation is given in [31]. The data from the mentioned work are used to form Table 2.12.

The crop management strategies concern: low-input integrated farming (LI), integrated farming following European Regulations (IFS), conventional farming (CONV). Marks in corresponding columns indicate operations that are conducted under a particular management system. It is evident that each of the management strategies leads to different energy consumptions. It is also claimed in [31] that the choice of the operation strategy does not affect the final energy obtained in the form of bioethanol obtained from the crop harvested from the plantation. The last opinion is to some extent contradictory to the existence of distribution of yields demonstrated in [27]. The probable explanation of the discrepancy is that opinion presented in [31], as well as similar opinions that appear in the literature, concerns particular production sites, and that the effects caused by the differences between crop management strategies applied to a particular site might be smaller than the effects caused by geographic factors that differentiate distant sites.

Similar studies on rapeseed planting technologies were published in [32]. Table 2.13 summarizes the operations performed in three tillage management systems described in the paper mentioned above. Table 2.14 presents energy consumption evaluated for the tillage management cases presented in previous Table 2.13.

The comparison of the data presented in Tables 2.13 and 2.14 indicates the major role of forecrops in determining energy consumption of plantation. The appropriate choice of the forecrop may, therefore, be a factor strongly affecting the performance of a plantation.

The recent studies performed in the frame of the grant from the European Commission gives insight into Energy consumption in agricultural operations in several European countries [33]. The average values of primary energy consumption in the cultivation of several crops are collected and compared. The data relevant for biofuels production are presented in Tables 2.15, 2.16, and 2.17.

It is seen that the amounts of energy required for agro-technical operations performed in the cultivation of the same plant in various countries appear quite different. The differences might be the result of differences in technology as well as the specific requirement of a particular soil, etc. Table 2.16 demonstrates that also yields of those plants differ in the countries discussed.

In consequence, the differences are seen in the specific primary energy consumption needed for the production of one metric ton of the crop.

These data indicate that for each production site, each equipment used, etc., individual estimation of energy consumption needed for the cultivation of a specific plant is required.

It should also be mentioned that planting some legume trees like pongamia pinnata, or genetically modified robinia pseudoacacia, acacia magnum, or leucaena leucocephala [34] can be regarded as future feedstock for biodiesel production.

Table 2.12 Energy consumed in subsequent agricultural operations of maize production in three production strategies

Month	Decade	Operation	Machine	Specific energy consumption	Crop management strategy		
				MJ/ha/a	LI	IFS	CONV
IV		Plowing	Moldboard—plough	1185.7		☑	☑
	2	Disking	Disk harrow	336.7	☑	☑	☑
		Disking	Disk harrow	322.1	☑		
		K fertilization	Centrifugal fertilizer spreader	91.5	☑	☑	☑
		Rototilling	Rotary harrow	900.3	☑	☑	☑
		N, *P* localized fertilization and planting	Row crop planter	245.2	☑		☑
		Row-localized *N* fertilization and preemergence weed control and planting	Row crop planter	245.2		☑	
	3	Preemergence weed control	Boom-type sprayer	120.8			☑
V	2	Post-emergence weed control	Boom-type sprayer	120.8	☑		
	3	*N* fertilization	Centrifugal fertilizer spreader	91.5	☑	☑	☑
VI	1	Row cultivation and furrowing	Row crop cultivator	322.1	☑	☑	☑
	2	Furrow irrigation			☑	☑	☑
VII	2	Furrow irrigation			☑	☑	☑
VIII	1	Furrow irrigation			☑	☑	☑
IX	2	Harvest	Self-propelled combine	1185.7	☑	☑	☑
		Straw shedding	Stalk shredder	596.5	☑	☑	☑
			Total energy consumed MJ/ha/a		4212.4	4955.2	5076

Table 2.13 Operations performed in different tillage management systems, and energy consumption for rapeseed plantations

Operations	Tillage management		
	Traditional	Simplified	Direct sowing
Forecrop	☑	☑	☑
Disking	☑		
Rototilling		☑	
First plowing	☑		
Harrowing	☑	☑	
Plowing	☑		
Rolling	☑		
Row-sowing		☑	☑
Sowing	☑		
Crop protection and fertilization	☑	☑	☑
Harvesting	☑	☑	☑

Table 2.14 Energy consumption in various variants of tillage management in rapeseed plantations

Tillage management:	Traditional	Simplified	Direct sowing
Operations	Energy consumption MJ/ha		
Without forecrop	1944	1404	1296
With root vegetable forecrop	6696	6156	6048
With papilionaceous forecrop	4169	3629	3521
With leguminous forecrop	3456	2916	2808

Table 2.15 Comparison of specific primary energy consumption for three plants in selected European countries in [GJ/ha]

Plant	Germany	Netherlands	Poland
Sugar beet	14	13.70	17.2
Potato	26.9	44.8	16.9
Wheat	18.6	18.1	15.1

Table 2.16 Comparison of yields of three plants in selected European countries in [t/ha]

Plant	Germany	Netherlands	Poland
Sugar beet	60.9	67.1	60
Potato	42.4	50.2	27
Wheat	7.66	8.73	5.8

Table 2.17 Comparison of specific primary energy consumption for three plants in selected European countries in [GJ/t]

Plant	Germany	Netherlands	Poland
Sugar beet	0.231	0.20	0.286
Potato	0.634	0.893	0.627
Wheat	2.4	2.1	2.6

2.5 Aquatic Biomass

The use of agricultural crops for biofuel production causes some doubts and controversies. Some works [35, 36] point to the problems that might occur due to the competition of biofuel crops with edible ones [37]. This competition might result in manifold reasons, and produce multiple results. The first is direct use of edible crops for industrial purposes, namely biofuels' production—which might result in a decrease of supply of food components, and consequently introducing an increase of the price of food. This increase would be especially perceptible by the poor part of the human population increasing the threat of hunger. The other problem is connected to indirect reasons, namely the elimination of the part of the arable land from food production, as well as putting additional load on water resources [38]. The positive aspects, being pointed out, are indicated as a decrease of energetic dependence on fossil fuel producers, as well as an increase of rural development. Some opinions suggest that the technological development in agricultural production [39] will permit coexistence and an increase in production of both: edible as well as energetic crops.

On the other hand, as a remedy for excessive land use, the energetic exploitation of biomass originated from the aquatic environment is postulated toward both food and biofuel production [40].

The solution that seems to be promising, maybe in both areas of the aquatic biomass use is the cultivation of biomass of microalgae. During recent years, numerous studies were dedicated to this topic leading to not always compatible results. The main direction of those studies is the investigation of biodiesel production from microalgae cultivated in artificial conditions. The simplified scheme of algae biofuel production system is presented in Fig. 2.8.

As indicated in [41], the individual processes shown in Fig. 2.8 are associated with several requirements, and may be realized in several versions. In the cultivation step, the presence of light, carbon dioxide, and appropriate nutrition, fresh or salt water, as well as correct temperature range (depending on algae strain) are essential factors. Several technical solutions are in use: the open pond is the simplest solution, while tubular or flat photo-bioreactors are more advanced. Since photosynthesis is the main process determining the growth of algae, the optimization of light utilization is of great importance. It is also associated with the phenomenon of light saturation—the too intense light causes a decrease in algae growth. In the closed reactors, the over-

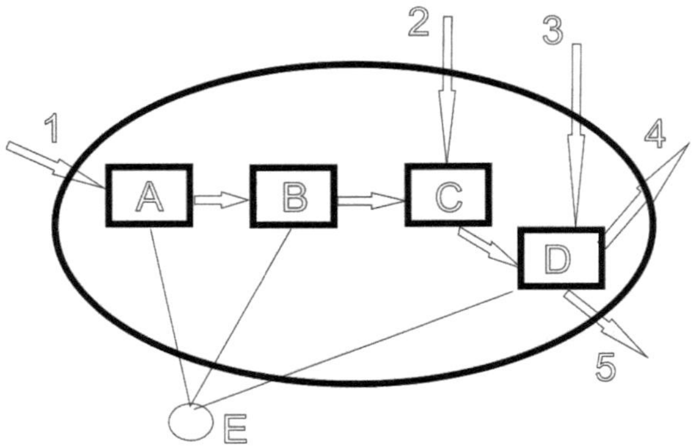

Fig. 2.8 Schematic representation of algae biofuel's production system. A—algae culture, B—dewatering, C—extraction, D—transesterification; inputs: 1—fertilizer, water, CO_2, 2—hexane, 3—methanol, H_3PO_4, KOH, E—Energy; outputs: 4—biodiesel, 5—glycerol, oil-cake

saturation with oxygen, produced by algae during photosynthesis, might introduce another problem leading to collapse of the algal colony.

Another critical factor, especially in closed systems exposed to solar radiation, is the temperature of the reactor, which, depending on algae strain, has to be kept in a relatively narrow range around 25 °C. Algal colonies are also sensitive toward pH of the surrounding, that, in turn, is affected by the presence of CO_2, the presence of nutrients, etc.

Open ponds do not show many of problems connected with the closed systems, but they are less productive, require a flow of larger volume of liquid, are subjected to evaporation of water, and also are sensitive to contamination.

The choice of algae strain is also important. Each of the available strains has different requirements toward cultivation conditions as well as very different productivities. According to the data presented in [42], the productivity of lipids may vary substantially. The values of lipids yield, expressed in mg per liter per day, are listed in Table 2.18. Various strains of algae also differ in the chemical composition of produced lipids: chemical individual containing a different number of carbon atoms are present in different proportions, and also different levels of unsaturation are observed [42]. Those molecular profiles reflect also in composition and properties of diesel fuels produced from different algae strains.

Harvesting of microalgae also requires special attention. The cultivation conditions require a very dilute suspension of algae, which strongly affects the separation of them from the water. The size of microorganisms is also a factor causing some difficulties. Several methods of separation are applied. Mechanical treatment usually precedes thermal drying processes. The mechanical dewatering may include flocculation, which consists of an increase of particle size by means of aggregation of individual microalgae. To achieve this, either multivalent cations are added, in order to neutralize the natural negative charge of the algae, or pH of the system

Table 2.18 Lipids productivity of various strains of algae

Algal species	Lipid productivity (mg L^{-1} day^{-1})
Kirchneriella lunaris	8.82
Selenastrum capricornutum	6.94
Staursatrum sp	0.84
Chlorella vulgaris	7.96
Scenedesmus obliqnus	6.57
Diatoms Navicula sp	3.67
Phaeodactylum tricornutum	26.75
Batrachospermum sirodotia	1.18
Lyngbya kuetzingii	7.75
Prymnesiophytes Isochrysis sphacrica	8.69

is changed by addition of acids or bases, depending on the algae strain. Aggregation of microalgae can also be caused by application of ultrasound. A further step is either sedimentation under forces of gravity, or centrifugal separation, which is faster than sedimentation, but requires more energy. Filtration is also considered as a method of algae separation from water. Both cake-type filters, and cross-flow filters are investigated, and some problems are found in each one. Flocculation preceding the filtration may increase the effectiveness of the process. Analysis of scientific and patent literature performed in [43] have identified several promising techniques, also suggesting a possibility of combining of some technological procedures.

The subsequent production step is the separation of microalgal oil from biomass. This goal also can be achieved by several routes. Mechanical methods consist of disruption of algae structure. Bead mills and presses are usual tools for this purpose. The correct choice of the type of mill or that of a press depends on the algal strain being cultivated and processed. In addition, extraction with some solvents has been proposed. This normally would be followed by distillation separating oil from the extraction solvent. One of the solvents could be the ready biodiesel that could not have to be removed prior to transesterification, and therefore would permit a decrease in energy consumption during processing. It was not considered, however, how the dilution of oil would affect the course of transesterification. As can be expected from chemistry, the presence of the reaction product in the reaction mixture affects the equilibrium as well as kinetics slowing the rate of reaction. Another publication [44] indicates possibilities for the production of hydrogen and ethanol besides of biodiesel from some algal cultures.

Recently, quite pessimistic, critical comments have been published [45]. This publication emphasizes that the big amount of money has been invested in the research, without positive results. Such conclusion, being in contradiction with known high photosynthetic efficiency, and early predictions of fast success, indicate that careful analysis of the present status of research followed by a search for new crucial attempts is necessary. On the other hand, the report published by the United Nations Food and Agriculture Organization [46] indicates the existence of several very innovative approaches, and solutions that are sufficiently promising.

2.6 Organic Wastes

Organic wastes originated in agriculture, food industry, etc., are the substances that require appropriate utilization. It seems quite natural that conversion of such wastes into biofuels should be the convenient way of utilization. Taking into account, that any utilization process requires some inputs of energy, even in the cases when no useful products or energy are created as a result, the utilization processes that give the output of energy can be considered useful. An example of various applications of dates' wastes for biofuel production (including biodiesel) is given in [47]. Another example is given in the paper [48]. The paper describes biodiesel production from chicken's fat by use of supercritical methanol method of transesterification. The paper does not report any data concerning energy inputs neither in logging of the fat wastes nor its processing into biodiesel. The other work [49] gives information about properties of biodiesel, indicating a generally higher cetane number of biodiesel originated from animal fat as compared to vegetable or seed oils. It also shows that different chemical compositions of fat obtained from different animals affects the chemical composition, and some properties of biodiesel produced from different types of fat. The mostly affected property is the solidification temperature. It was also shown [50] that the used oil from the restaurant can successfully be converted into biofuel. The data concerning energy inputs connected with the collection of wasted oils from restaurants, as well as conversion to biofuel are not reported. Life cycle analysis performed in [51] for the case of energetic conversion of municipal wastes shows that energy output from such installation is rather slightly smaller than energy inputs. Since the treatment of wastes is necessary from both social and ecological points of view such situation is acceptable.

References

1. Brostow, W., Datashvili, T., Miller, H.: Wood and wood derived materials. J. Mater. Educ. **32**(3–4), 125–138 (2010)
2. Belbo, H.: Technical specifications for solid biofuels. Evaluation of the new technical specifications provided by CEN/TS 335 in the Swedish Biofuel Market. SLU Master Thesis Nr 012 2006, ISSN 1651-761x
3. Krajnc, N.: Wood fuel handbook. Food and Agriculture Organization of United Nations, Pristina (2015)
4. Pong, W.Y., Waddell D.R., Lambert M.B.: Wood density-moisture profiles in old-growth Douglas-Fir and Western Hemlock, Res. Pap. PNW-347. US Department of Agriculture, Forest Service, Pacific Northwest Research Station Portland, OR (1986). 30 p

5. Terragni, O.: 80% of earth's forests have been destroyed; http://greenactionnews.net/blog/2014/10/05/80-of-earths-forests-have-been-destroyed-who-is-clear-cutting-the-most/. Visited 21.02.2018
6. Kellogg, L.D., Bettinger, P., Studier, D.D.: Terminology of ground-based mechanized logging in the Pacific Northwest. College of Forestry, Forest Research Laboratory, Oregon State University, Corvallis, OR (1993)
7. Okanda Oyier, P.: Fuel consumption of timber harvesting systems in New Zealand, A thesis submitted in partial fulfilment of the requirements for the degree of Master of Forestry Science at the School of Forestry University of Canterbury (2015 November)
8. Smidt, M., Gallagher, T.: Factors affecting fuel consumption and harvesting costs. Paper presented at the Council on Forest Engineering, Missoula, Montana (2013)
9. Lijewski, P., Merkisz, J., Fuć, P., Ziółkowski, A., Rymaniak, L., Kusiak, W.: Fuel consumption and exhaust emissions in the process of mechanized timber extraction and transport. Eur. J. Forest Res. **136**, 153–160 (2017)
10. Czemko, B.: Ile naprawdę mamy drewna w lesie? Rynek Drzewny 2/2007 How much wood we really have in the forest? Wood Market 2/2007
11. Matthews, R.W., Jenkins, T.A.R., Mackie, E.D., Dick, E.C.: Forest yield: a handbook on forest growth and yield tables for British forestry. Forestry Commission, Edinburgh (2016)
12. Christersson, L., Verma, K.: Short-rotation forestry—a complement to "conventional" forestry. Unasylva **223**(57), 34–39 (2006)
13. Eppler, U., Petersen, J.-E., Couturier, Ch.: Short rotation forestry, Short Rotation Coppice and perennial grasses in the European Union: agro-environmental aspects, present use and perspectives. In: Dallemand, J.F., Petersen, J.E., Karp A. (eds.) Short Rotation Forestry, Short Rotation Coppice and perennial grasses in Proceedings of the Expert Consultation: the European Union: Agro-environmental aspects, present use and perspectives. European Commission Joint Research Centre Institute for Energy JRC 47547 EUR 23569 EN ISSN 1018-5593, Harpenden, United Kingdom (2007 October 17 and 18)
14. Hinchee, M., Rottmann, W., Mullinax, L., Zhang, Ch., Chang, Sh., Cunningham, M., Pearson, L., Nehra N.: Short-rotation woody crops for bioenergy and biofuels applications. In: Tomes, D., Lakshmanan, P., Songstad D. (eds.) Biofuels, Global Impact on Renewable Energy, Production Agriculture, and Technological Advancements. Springer, New York, NY. ISBN 978-1-4419-7144-9
15. Pope, P.E., Dawson J. O.: Central Hardwood Notes Short-Rotation Plantations. In: Van Sambeek, J.W., Dawson, J.O., Ponder, F., Jr., Loewenstein, E.F., Fralish, J.S. (eds.) Proceedings, 13th Central Hardwood Forest conference; 2002 April 1–3; Urbana, IL. Gen. Tech. Rep. NC-234. St. Paul, MN: U.S. Department of Agriculture, Forest Service, North Central Research Station
16. Tubby, I., Armstrong A.: Management of short rotation coppice. Forestry Commission 231 Corstirphine Road, Edinburgh. ISBN 0-85538-567-7
17. Guidi, W., Pitre, F.E., Labrecque, M.: Short-rotation coppice of willows for the production of biomass in Eastern Canada. In: Matovic M.D. (ed.) Biomass Now. Sustainable Growth and Use. IntechOpen (2013)
18. Rédei, K., Csiha I., Keser Z.: Black locust (*Robinia pseudoacacia* L.) short-rotation crops under marginal site condition. Acta. Silv. Lign. Hung. **7**, 125–132 (2011)
19. Benoît, G., Maupu, P., Vial, E.: Life cycle assessment of eucalyptus short rotation coppices for bioenergy production in southern France. GCB Bioenergy **5**, 30–42 (2013). https://doi.org/10.1111/gcbb.12008
20. Dimitriou, I., Rutz, D.: Sustainable short rotation coppice. In: A Handbook. WIP Renewable Energies, Munich, Germany (2015)
21. Bacenetti, J., Fiala, M.: Short rotation coppice in Italy: a model to asses economic, energetic and environmental performances of different crop systems. World Renewable Energy Congress 2011—Sweden, Linkoping, Sweden (2011 May 8–13)
22. Gallagher, C., Murphy, J.D.: Is it better to produce biomethane via thermochemical or biological routes? An energy balance perspective. Biofuels, Bioprod. Bioref. **7**, 273–281 (2013)

23. Stolarski, M.J., Krzyżaniak, M., Szczukowski, S., Tworkowski, J.: Efektywność energetyczna produkcji biomasy wierzby w jednorocznym i trzyletnim cyklu zbioru, (Energetic effectiveness of willow biomass production in annual and triennial harvest cycle). Fragm. Agron. **31**(2), 88–95 (2014)
24. Wójcicki Z.: Metodyka badania energochłonności produkcji rolniczej. (Methodology of investigation of Energy consumption in agricultural production.) Problemy inżynierii rolniczej Problems of Agricultural Engineering. **4**(90) 17–29 (2015)
25. Pimentel, D.: Energy flows in agricultural and natural ecosystems. Options Mediterraneennes, pp. 125–136. Inst. Agron. Mediterraneo de Zaragoza, Zaragoza, Spain (1984)
26. Orynycz O., Chodorski A., Wasiak A.: Energetic efficiency of salix viminalis plantation. In: IX International Scientific Symposium "Farm Machinery and Processes Management in Sustainable Agriculture", Lublin, Poland (2017)
27. Johnston, M., Foley, J. A., Holloway, T., Kucharik, C., Monfreda, C.: Resetting global expectations from agricultural biofuels Environ. Res. Lett. 4 (2009) 014004 (9 pp.) https://doi.org/10.1088/1748-9326/4/1/014004
28. Johnston, M., Licker, R., Foley, J., Holloway, T., Mueller, N.D., Barford, C., Kucharik, C.: Closing the gap: global potential for increasing biofuel production through agricultural intensification. Environ. Res. Lett. **6**, 034028 (2011) (11 pp)
29. DeLucia, E.H., Gomez-Casanovas, N., Greenberg, J.A., Hudiburg, T.W., Kantola, I.B., Long, S.P., Miller, A.D., Ort, D.R., Parton, W.J.: The theoretical limit to plant productivity. Environ. Sci. Technol. **48**(16), 9471–9477 (2014). https://doi.org/10.1021/es502348e
30. Salla, D.A., Furlaneto, F.P.B., Cabello, C., Kanthack, R.A.D.: Análise energética de sistemas de produção de etanol de mandioca, (Energetic analysis of the ethanol production systems of cassava) (Manihot esculenta Crantz). Revista Brasileira de Engenharia Agrícola e Ambiental v.14, n.4, p. 444–448, 201
31. Alluvione, F., Moretti, B., Sacco, D., Grignani, C.: EUE (energy use efficiency) of cropping systems for a sustainable agriculture. Energy **36**, 4468–4481 (2011)
32. Orynycz, O., Wasiak, A.: Effects of tillage technology on energetic efficiency of rapeseed plantation for biofuel production. Appl. Comput. Sci. **10**(2), 67–76 (2014)
33. Gołaszewski, J., de Visser, Ch. (eds.): State of Art. On Energy Efficiency in Agriculture. Country data on energy consumption in different agroproduction sectors in the European countries. The project Agriculture and Energy Efficiency © 2012 agrEE, founded by the FP7 Program of the EU with the Grant Agreement Number 289139. Project Deliverable 2.1
34. Biswas, B., Scott, P.T., Gresshoff, P.M.: Tree legumes as feedstock for sustainable biofuel production: opportunities and challenges. J. Plant Physiol. **168**, 1877–1884 (2011)
35. Coyle, W.: The future of biofuels a global perspective, wcoyle@ers.usda.gov. Retrieved 12. 05.2018
36. Rosegrant, M.W., Msangi, S., Sulser, T., Valmonte-Santos, R.: Bioenergy and agriculture: Promises and challenges biofuels and the global food balance. Int. Food Policy Res. Inst.; Focus 14 • Brief 3 Of 12 • December 2006
37. Hill, J., Nelson, E., Tilman, D., Polasky, S., Tiffany, D.: Environmental, economic, and energetic costs and benefits of biodiesel and ethanol biofuels. Proc. Natl. Acad. Sci. U.S.A. **103**, 11206–11210 (2006)
38. de Fraiture, Ch., Giordano, M., Liao, Y.: Biofuels and implications for agricultural water use: blue impacts of green energy. Water Policy **10**(Supplement 1), 67–81 (2008)
39. https://www.eia.gov/conference/2014/pdf/presentations/calabotta.pdf. Retrieved 12. 05. 2018
40. Naylor, R.L.: Oil crops, aquaculture, and the rising role of demand: a fresh perspective on food security. Global Food Secur. **11**, 17–25 (2016)
41. Lassing, M., Mårtensson, P., Olsson, E., Svensson, M., Hulteberg, Ch., Karlsson, H.T., Lund University, Børresen B.T.,: StatoilHydro ASA, Eklund H., StatoilHydro ASA, KET050 Biodiesel Production from Microalgae Dept of Chemical Engineering, Lund University, Faculty of Engineering; Final Report on Biodiesel Production from Microalgae—A Feasibility Study—Presented to StatoilHydro ASA Oslo, Norway May 16, 2008

42. Song, M., Pei, H., Hua, W., Ma, G.: Evaluation of the potential of 10 microalgal strains for biodiesel production. Biores. Technol. **141**, 245–251 (2013)
43. Deconinck, N., Muylaert, K., Ivens, W., Vandamme, D.: Innovative harvesting processes for microalgae biomass production: A perspective from patent literature. Algal Res. **31**, 469–477 (2018)
44. Costa, A.V., Greque de Morais, M.: The role of biochemical engineering in the production of biofuels from microalgae;m. Biores. Technol. **102**, 2–9 (2011)
45. Wesoff, E.: Hard lessons from the great Algae biofuel bubble; https://greentechmedia.com/authors/eric-wesoff. April 19, 2017. Retrieved 28.05.2018
46. van Iersel, S., Flammini A.: Algae-based biofuels: applications and co-products FAO Environmental and Natural Resources Service Series, No. 44. FAO, Rome (2010)
47. Ramesh, D., Karthikeyan, S., Chinnanchetty, G.: Exploring the possibility of biofuel production from by-products ans wastes generated from date industry. In: Manickavasagan, A., Mohammed, M., Sukumar E.E. (eds.) Dates: Production, Processing, Food and Medicinal Values; CRC Press
48. Shi, W., Li, J., He, B., Yan, F., Cui, Z., Wu, K., Lin, L., Qian, X., Cheng, Y.: Biodiesel production from waste chicken fat with low free fatty acids by an integrated catalytic process of composite membrane and sodium methoxide. Bioresour. Technol. **139**, 316–322 (2013). https://doi.org/10.1016/j.biortech.2013.04.040. Epub 2013 Apr 19
49. Van Gerpen, J.: Animal fats for biodiesel production; farm energy January 31, 2014; http://articles.extension.org/pages/30256/animal-fats-for-biodiesel-production
50. Carmona-Cabello, M., Leiva-Candia, D., Castro-Cantarero, J.L., Pinzi, S., Dorado, M.P.: Valorization of food waste from restaurants by transesterification of the lipid fraction. Fuel **215**, 492–498 (2018)
51. Lausselet, C., Cherubini, F., del Alamo Serrano, G., Becidan, M., Strømman, A.H.: Life-cycle assessment of a waste-to-energy plant in central Norway: current situation and effects of changes in waste fraction composition. Waste Manag. **58**, 191–201 (2016)

Chapter 3
Technologies of Biofuel Production

3.1 Classification of Biofuels

Various technological processes have been developed in order to convert many types of biomass into some final energy vector (a biofuel). Those final biofuels differ in the physical form, being produced as liquids, gases, or solid substances. The methane, and recently hydrogen, derived from biological resources, can be given as the examples of gas-type biofuels, while biodiesel or bioethanol, and some higher hydrocarbons exemplify liquid biofuels. Solid biofuels, in turn, occur in the forms of chopped wood pieces, in the most classical form, and chips, pellets, or briquettes made either of wood chips, straw, etc. The biofuels also differ in chemical composition as well as in the ways of application, e.g., for heat generation, electricity generation, automotive fuels, etc. The variety of the biological resources, as well as diversification of the biofuels generate multiple ways of conversion of these resources into fuels.

Biofuels are usually classified as primary and secondary ones. The primary biofuels include fuelwood, wood chips, pellets, etc. These fuels are used in an unprocessed, non-modified chemical form, mainly being directly combusted for heating or electricity production. The secondary biofuels are produced from primary ones in processes converting biomass into solid (e.g., charcoal or torrefied pellets), liquid (e.g., ethanol, biodiesel, DME, etc.) or gases (e.g., biogas, synthesis gas, and hydrogen). Those fuels are used for powering vehicles and can be used in various industrial processes. The secondary biofuels are, in turn, divided into three generations. This classification is based upon the raw material and technology used for the production of particular fuel.

Biofuels can also be classified according to their origin. They may be produced from biomass extracted from forests, agriculture, fishery as well as from municipal or industrial wastes (especially from the food processing industry). A variety of processing technologies are already elaborated offering numerous types of biofuels, and numerous pathways for their production [1, 2]. Separate technologies of biofuel production were elaborated having in mind either the resource from which the process

A. Wasiak, *Modeling Energetic Efficiency of Biofuels Production*, Green Energy and Technology, https://doi.org/10.1007/978-3-319-98431-5_3

is started or the final result, i.e., the form, in which a biofuel is obtained. It may concern just physical appearance or chemical composition. From the physical point of view, three forms—gaseous, liquid, or solid—can be obtained. Obtaining each of the forms of biofuels requires special technological approach.

3.2 Energetic Aspects of Biofuels Production

Gaseous fuels are obtained either by gasification processes of chemical nature [3], or by biochemical processes like anaerobic digestion [4]. The biogas production is frequently combined either directly to a gas-fueled electricity generator or is connected to gas distribution system pipes. Various pretreatment methods [5] were also analyzed for the case of biogas production from the grass.

Recently several production processes are designed as hybrid ones simultaneously generating gas and a liquid fuel [4]. A comparison of similar processes has been given in [6]. Some data from the work [6] are listed in Table 3.1 showing ethanol and methane yield from such a hybrid process, and the total energy yield for corresponding methods of treatment. Unfortunately, the amount of energy needed to carry out such processes is not reported.

Since liquid biofuels have a very wide range of application, they are also intensively developed, produced, and investigated, mainly in order to look for better performance or economic feasibility. The simplest technologies concern s.c. first generation of liquid biofuels. The main example of first-generation biofuels is bioethanol, which production by fermentation of sugars has a very long tradition although not for this application. Later, the methods of fermentation included conversion of starch

Table 3.1 Products and energy yield in hybrid processes of simultaneous generation of bioethanol and biomethane

Substrate	Pretreatment	Ethanol/methane [g/kg substrate]	Total energy yield [MJ/kg]
Birch	Steam exploded	166/86	9.13
Corn stover	Soaking	127/64	6.96
Corn stover	Steam exploded	150/99	9.53
Corn stover	Steam exploded	197/96	10.62
Hemp	Steam exploded	48/130	11.2
Oat straw	Steam exploded	150/91	9.08
Sugarcane bagasse	Peroxide	201/49	8.11
Sugarcane bagasse	Steam exploded	137/54.6	7.11
Sugarcane bagasse	Alkali (0,5 NaOH)	162/142	12.25

Source Own elaboration on the basis of [6]

Table 3.2 Production of main biofuels during the year 2008

Country	Fuel	Gross production [1]	EROEI	Net energy [EJ]
US	Ethanol	38×10^9	1.25	0.168
Brazil		26.5×10^9	5	0.469
others		26.5×10^9	5	0.469
UE	Biodiesel	5.9×10^9	1.5	0.065
other		5.9×10^9	3	0.13

Source Own elaboration based on [12]

Table 3.3 EROEI of biofuels produced in some countries

Biofuel	Country	EROEI
Ethanol from sugarcane	–	0.8–10
Ethanol corn-based	US	0.8–1.6
Biodiesel	US	1.3
Fossil fuels	–	10–80

Source Own elaboration on the basis of [13]

[7], and more recently lignocellulosic biomass, what is already classified as a second generation of biofuels.

The other example of first-generation biofuels is biodiesel produced by transesterification of vegetable oils. Several factors are indicated as the disadvantages of this type of product. At first, only a very small fraction of oleaginous plants biomass [8] (i.e., the grains) is used for conversion to biofuel, the rest is almost useless residue. The second argument is raised with respect to the use edible oils as a resource for industrial purposes, what might negatively affect the food market. Several contradictory opinions concerning this problem are published [9–11].

Production of biofuels has already reached substantial share in the world's economy. Some insight into the size of production can be obtained from the data [12] contained in Table 3.2. The EROEI indicator[1] given in Table 3.2 is quite different for different countries, and both fuels. Since EROEI > 1 in all the cases shown it means that in the biofuel production some energy gain is achieved. In the case of biofuels, this is an enormously important factor because the idea of biofuel production is to replace fossil fuels. The replacement (i.e., energy contained in the biofuel) cannot be smaller than the input of energy needed for production. It seems, however, that not in all cases this condition is fulfilled. The data given by [13], summarized in Table 3.3 show that in some cases the condition mentioned above is not fulfilled.

The work [14] reports improvement achieved in corn-based ethanol production resulting in significant increase of net energy balance from 1.26 in the old process to 1.76 in the new one.

[1] A close description of the EROEI is given in Sect. 4.2.

Table 3.4 Energy consumption in biodiesel production

Input	Quantity	Unit	Energy Intensity	Unit	Energy cost (MJ/ha)
Oil pressing					
Electricity	554.1	MJ	3.8	MJ/MJ	2105.6
Crude oil purification					
Electricity	29.7	MJ	3.8	MJ/MJ	112.9
Transesterification					
Electricity	377.8	MJ	3.8	MJ/MJ	1435.6
Methanol	78.4	kg	54.0	MJ/kg	4233.6
Sodium hydroxide	7.3	kg	72.6	MJ/kg	529.9
Hydrochloric acid	26.5	kg	23.5	MJ/kg	622.7
Water	81.9	t	3.5	MJ/t	286.6
Total					9327.1

Table 3.5 Energy requirement for oil extraction from canola and soybean

Inputs	Energy input	Canola	Soybean
	MJ/Mg	MJ/L	MJ/L
Crushing	1037	3.11	6.90
Feedstock transfer	107	0.32	0.71
Process heat	160	0.48	1.07
Conditioning	16	0.05	0.11
Total	1320	3.96	8.79

Source Own elaboration on the basis of [16]

The work [15] reports the energy consumption in biodiesel production. The energy consumption in several operations connected with the industrial conversion of oil into biodiesel is presented in Table 3.4. The reported energy cost is referred to 1 hectare of rapeseed plantation. The authors of the work also claim that the total energy cost is 1.1 higher than total energy output, so the net energy output is negative. Evidently, such a production system needs improvement. Such a result also poses the question of economical nature, i.e., whether or not the negative energy balance is accompanied with a negative financial balance? Frequently, the financial losses are compensated from governmental support, what prevents producers from looking for better technological solutions.

Some papers give reports only for particular parts of the industrial process. An example is the work [16]. Every such work is valuable, because it helps to build imagination about the energy requirements of particular processes, and enables looking for eventual improvements (Table 3.5).

Table 3.6 Energy consumption in the process of wood pellets torrefaction

Torrefaction conditions		250–30	270–30	270–60	290–30	Wood Pellet
Electricity	MJ/Mg	12.10	12.10	12.10	12.10	12.10
Utility fuel	MJ/Mg	1530.60	1278.90	407.20	548.20	1134.20
Total I	MJ/Mg	1542.70	1291.00	419.30	560.30	1146.30
Pellet production						
Electricity	MJ/Mg	342.00	342.00	342.00	342.00	530.00
Total II	MJ/Mg	1884.70	1633.00	761.30	902.30	1676.30

It has to be mentioned that also the production of solid fuels, especially those belonging to the second generation requires expenditures of energy. As an example, Table 3.6 gives a comparison of energy consumption in several conditions of wood pellets torrefaction. Unfortunately, this work does not report calorific values of pellets torrefied in different conditions.

Summarizing this chapter, it seems worth to point that in spite of a large amount of publications concerning biofuels the reliable data on energy consumption during biofuels production are not easily available. Some producers consider them as an important secret.

It indicates that mathematical modeling helps in building insight into interactions between different characteristics of the production system which may be really help in proposing improvements of the energetic productivity of such systems.

References

1. Thrän, D., Dotzauer, M., Lenz, V., Liebetrau, J., Ortwein, A.: Flexible bioenergy supply for balancing fluctuating renewables in the heat and power sector—a review of technologies and concepts. Energy, Sustain. Soc. **5**(35), 1–15 (2015)
2. Nigam, P.S., Singh, A.: production of liquid biofuels from renewable resources. Prog. Energy Combust. Sci. **37**, 52–68 (2011)
3. Sharma, A., Terrell, E., Theegala, C.S.: Biomass gasification and physical analysis of plant biomass and agricultural waste products in Louisiana. Wood Fiber Sci. **49**(3), 1–11 (2017)
4. Valli, L., Rossi, L., Fabbri, C., Sibilla, F., Gattoni, P., Dale, B.E., Kim, S., Ong, R.G., Bozzetto, S.: Greenhouse gas emissions of electricity and biomethane produced using the Biogasdoneright™ system: four case studies from Italy. Biofuels Bioprod. Bioref. **11**, 847–860 (2017). https://doi.org/10.1002/bbb.178
5. Rodriguez, C., Alaswad, A., Benyounis, K.Y., Olabi, A.G.: Pretreatment techniques used in biogas production from grass. Renew. Sustain. Energy Rev. (2016). https://doi.org/10.1016/j.rser.2016.02.022i
6. Kalyani, D.C., Zamanzadeh, M., Müller, G., Horn, S.: J: Biofuel production from birch wood by combining high solid loading simultaneous saccharification and fermentation and anaerobic digestion. Appl. Energy **193**, 210–219 (2017)

7. Larson, E.D.: Biofuel production technologies: status, prospects and implications for trade and development. Report No. UNCTAD/DITC/TED/2007/10. United Nations Conference on Trade and Development, New York and Geneva (2008)
8. Escobar, J.C., Lora, E.S., Venturini, O.J., Yanez, E.E., Castillo, E.F., Almazan, O.: Biofuels: environment, technology and food security. Renew. Sustain. Energy Rev. **13**, 1275–1287 (2009)
9. Van Dijk, M., Meijerink, G.W.: A review of global food security scenario and assessment studies: Results, gaps and research priorities. Global Food Secur. **3**, 227–238 (2014)
10. Naylor, R.L.: Oil crops, aquaculture, and the rising role of demand: a fresh perspective on food security. Global Food Secur. **11**, 17–25 (2016)
11. Naylor, R.L.: Higgins M. M.: The rise in global biodiesel production: Implications for food security. Global Food Secur. **16**, 75–84 (2018)
12. de Castro, C., Carpintero, Ó., Frechoso, F., Mediavilla, M., de Miguel, L.J.: A top-down approach to assess physical and ecological limits of biofuels. Energy **64**, 506–512 (2014)
13. Hall, C.A.S., Lambert, J.G., Balogh, S.B.: EROI of different fuels and the implications for society. Energy Policy **64**, 141–152 (2014)
14. Wang, K., Zhang, J., Tang, L., Zhang, H., Zhang, G., Yang, X., Liu, P., Mao, Z.: Establishment and assessment of a novel cleaner production process of corn grain fuel ethanol. Biores. Technol. **148**, 453–460 (2013)
15. Chen, H., Chen, G.Q.: Energy cost of rapeseed-based biodiesel as alternative energy in China. Renewable Energy **36**, 1374–1378 (2011)
16. Fore, S.R., Porter, P., Lazarus, W.: Net energy balance of small-scale on-farm biodiesel production from canola and soybean. Biomass Bioenerg. **35**, 2234–2244 (2011)

Chapter 4
Energetic Efficiency of Biofuel Production

4.1 Thermal Properties of Biofuels

Both primary and secondary biofuels have to assure specified requirements concerning their properties. Since the biofuels can exist in solid, liquid, and gaseous forms, some of the characteristics are very different, some are universal for all types of fuels. The variety of properties for technical, as well as market purposes are carefully standardized [1].

The most important property is calorific value (heating value). As shown in numerous research the calorific value results of chemical composition of a fuel. Many formulas have been proposed to describe the correlation between heating value, and elemental or on proximate data on the fuel composition. Two main cases of calorific value are distinguished: the higher heating value (also called gross calorific value) and low heating value (also called net calorific value). Those calorific values differ because the majority of fuels contain hydrocarbons, which causes the combustion gases to contain water. Evaporation of this water consumes some part (dependent upon hydrogen content) of energy released during combustion of the fuel. High calorific value is determined in such a way that the corresponding amount of latent heat of evaporation is not subtracted from reaction heat. Low calorific value, in turn, corresponds to this amount of energy, which remains after evaporation of water, and is available to the user. Another contribution of the same type comes from moisture—the molecules of water absorber by the fuel.

The methods of determination of both calorific values as well as their precise definitions are given in appropriate standards (e.g., ISO1213-2:1992) [2].

Correlation between the chemical composition of a fuel and its calorific value was first studied by Dulong, whose historical equation can be written in the following form [3]:

A. Wasiak, *Modeling Energetic Efficiency of Biofuels Production*, Green Energy and Technology, https://doi.org/10.1007/978-3-319-98431-5_4

$$Q = 0.336\text{C} + 1.418\text{H} + 0.094\text{S} - 0.145\text{O} \quad (4.1)$$

where

Q higher calorific value in MJ/kg
C mass% content of carbon in the fuel
H mass% content of hydrogen
S mass% content of sulfur
O mass% content of oxygen.

As the other example, the formula proposed by Lloyd and Davenport [4] can be given. This assumes the following form:

$$Q = 0.3578\text{C} + 1.1357\text{H} + 0.059\text{N} + 0.1119\text{S} - 0.0845\text{O} \quad (4.2)$$

where the meaning of symbols is the same as in the former case.

The recent studies, [5] performed on a number of wood species, other plants, as well as model materials being a mixture of pure components (cellulose, lignin, etc.) brought authors to a conclusion that the two following equations are the best fitting of experimental data:

$$Q = 0.00355\text{C}^2 - 0.232\text{C} - 2.230\text{H} + 0.0512\text{CH} + 0.131\text{N} + 20.600 \quad (4.3)$$

or

$$Q = 0.328\text{C} + 1.4306\text{H} - 0.237\text{N} + 0.0929\text{S} - \left(1 - \frac{\text{Ash}}{100}\frac{40.11\text{H}}{\text{C}}\right) \quad (4.4)$$

where C, H, N, S—are concentrations (in mass%) of corresponding elements, while Ash is a content of ash. The equations are applicable to dry matter.

It should be also mentioned that Eq. (4.3) was a few years earlier proposed in the paper of other authors [6] as a result of fitting equations of various form to the results of experimental measurements for biomass samples.

Another idea is represented in the work, where the calorific value of the biomass is correlated with the contents of chemical substance, estimated on the basis of proximate analysis (Table 4.1).

Table 4.1 Measured values of lignin content and higher heating value of some biomass samples

Biomass sample	L [wt%]	HHV [MJ/kg]
Corn stover	14.4	17.8
Corncob	15.0	17.0
Sunflower shell	17.0	18.0
Beechwood	21.9	19.2
Ailanthus wood	26.2	19.0
Hazelnut shell	42.5	20.2
Wood bark	43.8	20.5
Olive husk	48.4	20.9
Walnut shell	52.3	21.6

Data from Ref. [7]

It was shown in the work [7] that HHV well correlates with the content of lignin leading to the formula.

$$\mathrm{HHV} = 0.09979L + 16.292 \tag{4.5}$$

where L—is weight percent content of lignin in the biomass sample.

4.1.1 Energetic Efficiency Indicators

Following the early qualitative attempts [cf. 8], being philosophical or economic nature, the energetic efficiency of the fuel production procedures were quantitatively defined in [9]. To that indicator, a name EROEI (or EROI) "Energy Return on Energy Invested" was given. It was defined as a ratio of energy delivered by the production system, being under consideration, to the energy that has to be introduced to that system in order to facilitate processes needed to obtain and deliver the above-mentioned amount of energy from particular resource, i.e.:

$$\mathrm{EROEI} = \frac{E_{\mathrm{delivered}}}{E_{\mathrm{input}}} \tag{4.6}$$

The E_{input} is usually the sum of all energy fluxes that are introduced to the system during various operations. EROEI can be used to estimate the efficiency of any type of energy harvesting system, which requires inputs of some amount of energy.

Obviously, this indicator may depend not only upon the types of processes involved in the conversion of the resource to the energy delivered, but also it may be a function of time. This time dependence might reflect changes in production environment, and the status of production technology. The boundaries of the EROEI can be easily evaluated, the lower limit is reached when no energy is delivered

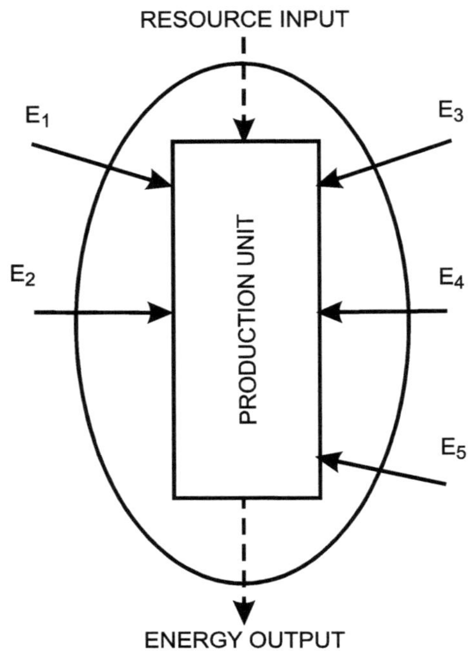

Fig. 4.1 A scheme of the fuel production system

($E_{\text{deliv}} = 0$)—then EROEI is also zero. The upper limit, in turn, is the case when no input of energy is required, then the EROEI reaches infinity.

Despite that EROEI is a widely accepted indicator, some disadvantages are discussed. Mostly the lack of precisely defined calculation procedures is indicated. As it is shown in [10], several computation procedures, and several definitions are used, leading to ambiguities in results [11, 12].

The idea of EROEI is represented in Fig. 4.1. The input to the production system is, first of all, a material resource that is used as the energy vector. The other inputs are energy fluxes, E_i, that facilitate the conversion of that resource into the fuel, and finally into energy. They are not directly converted into energy.

Some other indicators of performance of the energy production systems have also been introduced [cf. 13, 14]. The simplest choice is a NEG (the net energy gain), which is defined as a difference between the energy delivered and energy consumed in the system.

$$\text{NEG} = E_{\text{delivered}} - E_{\text{input}} \tag{4.7}$$

It is easy to recognize that NEG is expressed in the units of energy, while EROEI is dimensionless. A similar approach is used in s.c. NEA methodology (Net Energy Analysis) [15–17], which is mainly used in economic and ecological considerations.

It should also be pointed out that in both indicators, described above, the amount of energy contained in the main energy vector undergoing conversion is not taken

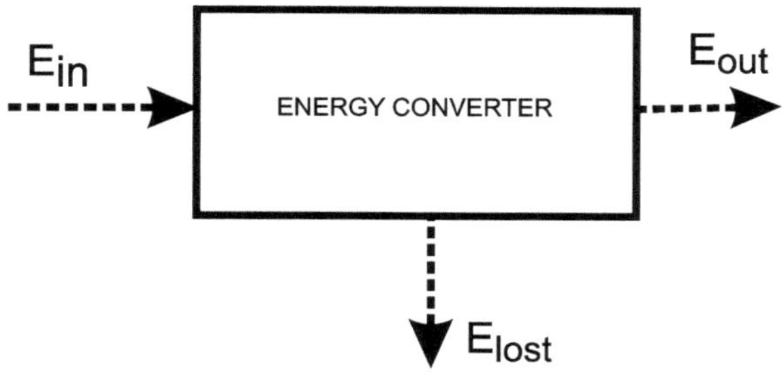

Fig. 4.2 Energy converter (the arrows show energy introduced to the converter, E_{in}, the energy delivered out the converter, E_{out}, and the energy lost in it E_{lost})

into consideration. Only the inputs of energy needed to facilitate conversion are compared to the energy delivered by the system.

In contrast, the thermodynamic energy efficiency of an energy converter is defined as follows:

$$\eta = \frac{E_{\text{out}}}{E_{\text{in}}} = \frac{E_{\text{out}}}{E_{\text{out}} + E_{\text{lost}}} \tag{4.8}$$

The meaning of symbols is explained in Fig. 4.2.

This quantity varies in the range $0 < \eta < 1$ since the smallest value of E_{out} may be zero, and then, η, is also zero, and on the other hand when the lost energy E_{lost}, would be zero than η reaches maximum equal to one.

In addition, when several converters are connected in series, it is easy to prove, that result can be expressed as a product of the values, η_n, for all the converters connected in the chain:

$$\eta_N = \prod_{1}^{N} \eta_n \tag{4.9}$$

where N, is a number of converters in the chain.

It will be demonstrated later that such a rule does not hold for the EROEI.

It is also important that it is extremely difficult to estimate the real amount of solar energy absorbed by a specific amount of biomass, and, therefore, to evaluate the real thermodynamic efficiency achieved in biofuel production.

A number of additional energy ratios, including NER—net energy ratio, NEER—net external energy ratio, GER—gross energy ratio have also been introduced [18]. These indicators may help to distinguish between efficiencies of processes of harvesting and conversion of energy that require external energy from those that can be fueled internally. Since the flow of energy through the system is explicitly considered, application of those indicators to energy derived from biomass might be difficult or even problematic, due to mentioned earlier limitations concerning the starting amount of solar energy absorbed by growing biomass.

Biomass can be considered as a type of energy storage that converts solar energy into material compounds, which later can be converted into energy. The relation

Table 4.2 Solar energy inputs needed for the formation of plant compounds compared to the higher heating values of those compounds estimated on the basis of chemical composition, and thermodynamic energy efficiency η

	Chemical formula	Solar energy input	HHV Ll & D	HHV Friedl	$\eta = \frac{HHV_{\mathrm{Friedl}}}{\mathrm{input}}$
		MJ/kg	MJ/kg	MJ/kg	
Citric acid	$C_6H_8O_7$	207.7		18.04	0.086
Monomer	$C_6H_{12}O_6$	261.9	15.96	15.76	0.06
Sucrose	$C_{12}H_{22}O_{11}$	271.6	16.65	16.63	0.061
Cellulose	$C_6H_{10}O_5$	282.3	22.59	15.63	0.055
Hemicellulose	$C_{11}H_{18}O_9$	284.3	17.57	17.75	0.062
Lignin coniferyl alcohol	$C_{10}H_{12}O_3$	429.7	27.74	28.83	0.067
Lipid glyceryl trioleate	$C_{57}H_{101}O_6$	553.3	37.42	44.06	0.079
Protein zein	$C_{4.6}H_{7.0}N_{1.0}O_{1.4}S_{0.02}$	1016.8	23.73	24.9	0.024

Data partially from Ref. [19]

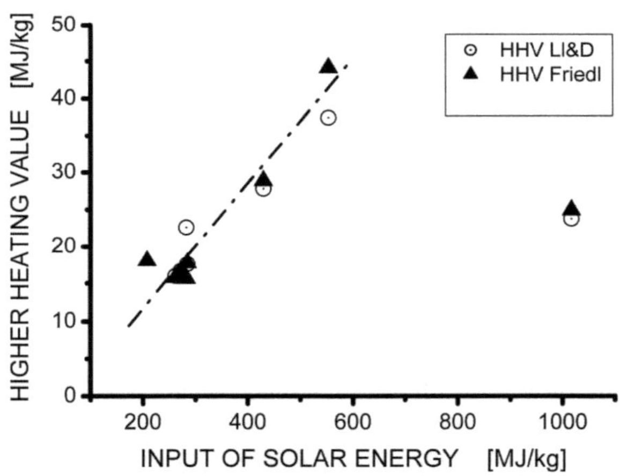

Fig. 4.3 Correlation between higher heating values of compounds listed in Table 4.2, and solar energy needed for their formation. The points for protein are far apart from the correlation line

between solar energy absorbed by the plant and the energy contained in the chemical components of the plant have been analyzed in [19]. The higher heating values (HHV) of the compounds relevant to the study, recalculated by the present author using Eqs. (4.2) and (4.3) are given in Table 4.2. The correlation between the higher heating value and the amount of solar energy needed to form a particular compound is shown in Fig. 4.3.

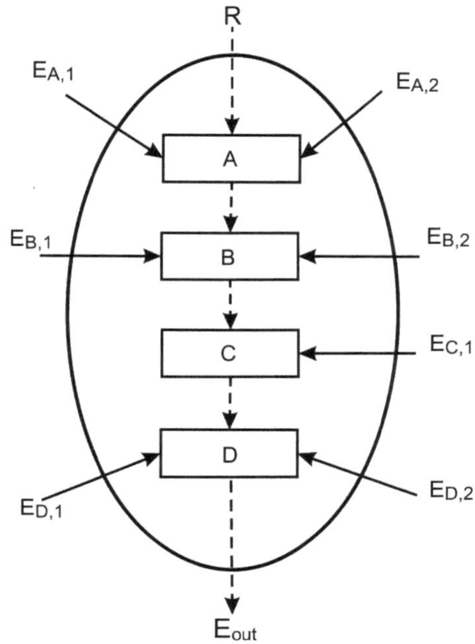

Fig. 4.4 The scheme of the biofuel production system containing several subsystems requiring inputs of energy for their operations

As seen in Fig. 4.3 a linear correlation is visible for all compounds, except protein, which requires much more solar energy to be formed, and still has a relatively low HHV (higher heating value) than the majority of remaining compounds. It may be concluded that biological species forming the high content of simpler compounds are more suitable as energy sources than ones forming the high content of proteins.

4.1.2 Partial Indicators for Subsystems Versus Total Efficiency of the System

Investigations of the effects caused by properties of individual parts of a production system on its energetic efficiency are the main idea of the approach of structural modeling of production systems. This idea requires the appropriate definition of indicators being used, as well as the determination of their properties, and establishing the rules of their additivity. From the viewpoint of application toward the production of biofuels, the EROEI appears to be the best starting point.

Analyzing Eq. (4.6) with respect to a system built of a series of subsystems, (or a series of operations), each one of those receives some inputs of energy as presented in Fig. 4.4 one can write an expression for energetic effectiveness as follows:

$$\varepsilon = \frac{E_{\text{out}}}{\sum_i E_{\text{in}_i}} \tag{4.10}$$

where E_{in_i} corresponds to the all individual fluxes of energy as it is exemplified in the figure.

Partial energetic effectiveness [20], for the individual subsystem, can be defined as

$$\varepsilon_A = \frac{E_{\text{out}}}{\sum_i E_{A,i}} \tag{4.11}$$

where E_{out} is the amount of energy delivered by the whole system, and $E_{A,i}$ are the energy inputs to the particular subsystem (in Eq. 4.11, it is the subsystem A—shown in the Fig. 4.4).

Obviously, such a partial indicator of energetic effectiveness can be defined for any subsystem constituting a given production system. Such definition enables derivation of a simple additivity rule [20, 21].

Namely

$$\frac{1}{\varepsilon} = \frac{\sum_N \sum_i E_{N,i}}{E_{\text{out}}} \tag{4.12}$$

where N numbers subsystems and i—the energy fluxes received by each of those subsystems, correspondingly.

Basing on Eqs. (4.11) and (4.12), the relationship between partial and global indicators of energetic effectiveness can obviously be written in the following form:

$$\frac{1}{\varepsilon} = \sum_N \frac{1}{\varepsilon_N} \tag{4.13}$$

What can be verbally expressed as: the global energetic effectiveness of a system is equal to the sum of the values of partial energetic effectiveness estimated for each subsystem, provided that all the partial indices are related to the same value of final energy delivered by the system investigated.

This relation holds for various structures of production systems.

For example energetic efficiency of a system shown in Fig. 4.5 that contains two production lines, that jointly deliver final energy can be expressed by Eq. (4.12). It has to be pointed out, however, that if computation of separate values of the effectiveness for each of the line is attempted, there are no simple relations of such values to the global one.

Another example can be given for the case when the system contains an energy converter that does not need to be externally powered (there is no additional input of energy), but loses some amount of energy, e.g., due to radiation. Such a case might occur when the biomass production system is connected to an electricity generator. Such a system is represented in Fig. 4.6.

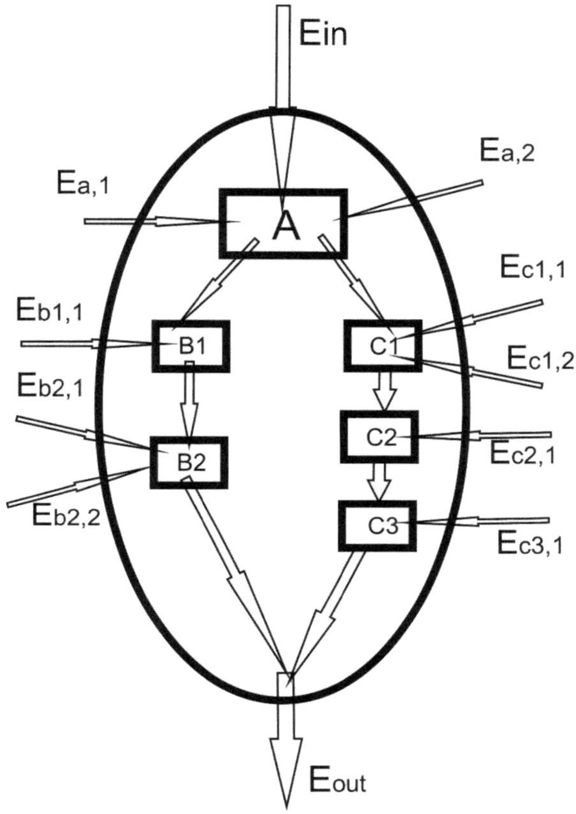

Fig. 4.5 The scheme of the biofuel production system containing several subsystems containing two lines of subsystems, finally giving a joint output of energy

In such a situation, the energy delivered is decreased by a factor of, η, which is defined in Eq. (4.8). Therefore, if the energy $E_{\text{bio},}$ delivered by the biological system is known, then effectiveness of the total system can be written as

$$\varepsilon = \frac{\eta E_{\text{bio}}}{\sum_i E_{\text{in},i}} = \eta \varepsilon_{\text{bio}} \tag{4.14}$$

where η is thermodynamic efficiency of the converter (D—in Fig. 4.6.), and ε_{bio} denotes energetic efficiency of the biological system alone.

Introduction of the idea of the partial energetic effectiveness of a subsystem is very useful for the structural modeling, because enables studying the influences of changes occurring in particular subsystems or particular operations on the energetic effectiveness of the whole system. The behavior of the energetic effectiveness of a complex system composed of several subsystems has been studied in [21]. Results of computations simulating the effect of the composition of two components are shown in Fig. 4.7.

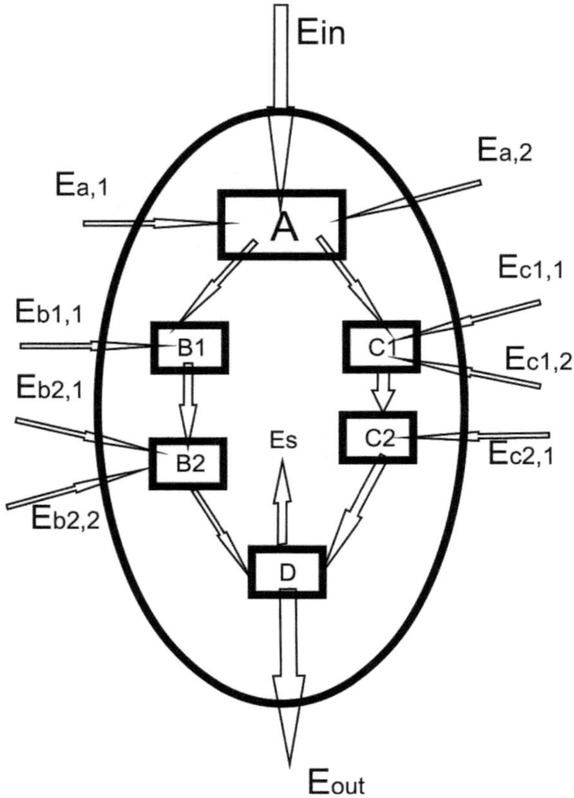

Fig. 4.6 Schematic representation of a biofuel production system that contains the converted generating some loses

For the sake of better visualization of the dependency selected values computed on the basis of Eq. (4.13) are collected in Table 4.3.

It can be seen that for the two-component system when the contributing values are equal, the global value is equal to half of the contributing ones. For the cases when contributing values are different, the resulting global value is smaller than the

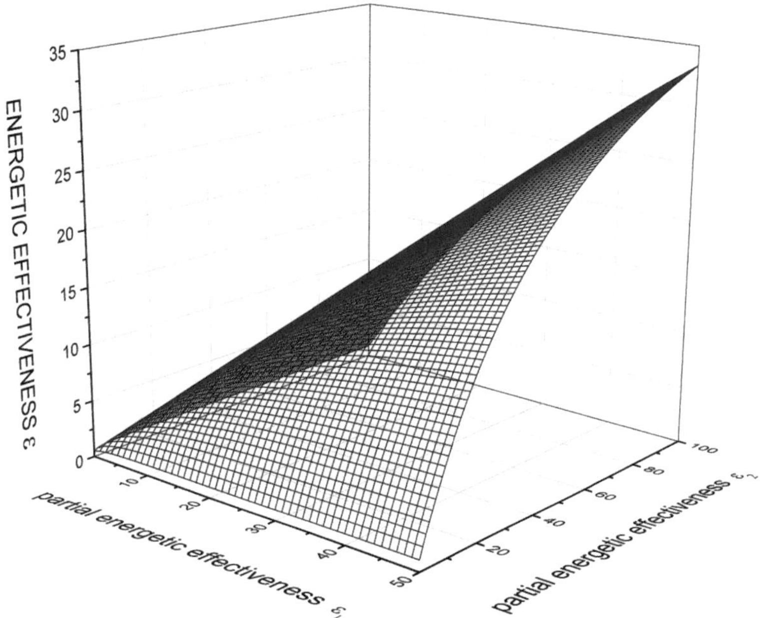

Fig. 4.7 Energetic effectiveness of a production system composed of two components with various values of the partial effectiveness of subsystems

Table 4.3 Dependence of the energetic effectiveness of a system upon values of contributing partial energetic effectiveness of subsystems

ε_2	10	20	30	40	50
ε_1					
10	5	6.7	7.5	8	8.4
20	6.7	10	12	13.4	14.3
30	7.5	12	15	17.2	18.8
40	8	13.4	17.2	20	22.3
50	8.4	14.3	18.8	22.3	25
60	8.6	15	20	24	27.3
70	8.8	15.6	21	25.5	29.2
80	8.9	16	21.9	26.7	30.8
90	9	16.4	22.5	27.7	32.2
100	9.1	16.7	23.1	28.6	33.4

smaller contributing partial value. Consequently, it has to be taken into account that the smaller contributing value mostly determines the global energetic effectiveness of the whole system. The observed effect becomes even more pronounced when the number of contributing subsystems increases. Table 4.4 shows the values for a three-

Table 4.4 Energetic efficiency of a three-component system in which $\varepsilon_3 = 50$, and ε_1 and ε_2 are equal to those listed in Table 4.3

ε_2	10	20	30	40	50
ε_1					
10	4.5	5.8	6.5	6.8	7.1
20	5.8	8.3	9.6	10.5	11.1
30	6.5	9.6	11.5	12.7	13.6
40	6.8	10.5	12.7	14.2	15.3
50	7.1	11.1	13.6	15.3	16.6
60	7.3	11.5	14.2	16.2	17.6
70	7.4	11.8	14.7	16.8	18.4
80	7.5	12.1	15.1	17.3	19
90	7.6	12.3	15.5	17.8	19.5
100	7.6	12.5	15.7	18.1	20

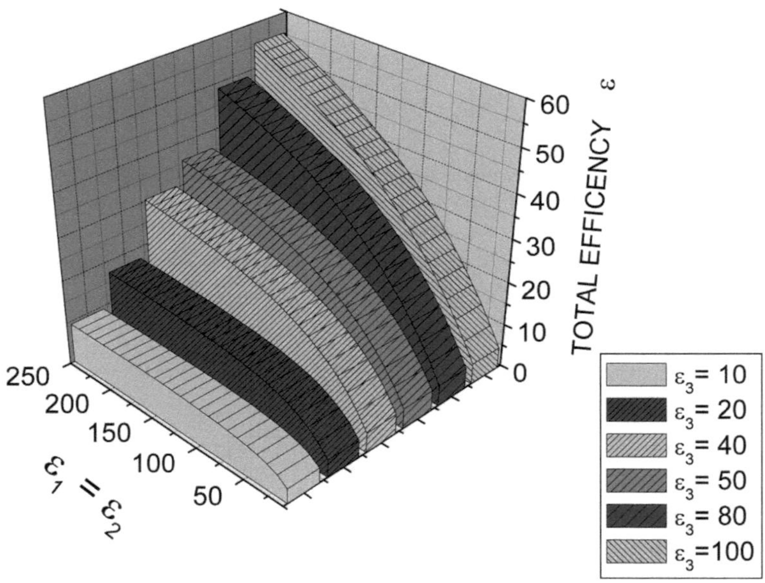

Fig. 4.8 The total energetic efficiency resulting from the combination of three subsystems differing in partial efficiency

component system, in which the third component has the value $\varepsilon_3 = 50$, and values of ε_1 and ε_2 are the same as in Table 4.3. The decrease of the efficiency is evident, in spite that the third component exhibits a quite high value of partial efficiency. These observations are of great importance for the design of the real biofuel production systems, as well as for any other fuel production systems.

The relations shown in Table 4.3 are also illustrated in Fig. 4.8.

It even more evidently shows how strong effect is associated to the combination of several subsystems differing in partial energetic efficiency.

The general conclusion that can be drawn from the above considerations is that in any real biofuel production system, the energetic effectiveness of an individual subsystem (or individual technological operation) should be as high as possible, and the number of subsystems (or operations) involved, should be as small as possible.

This conclusion can be considered as a guideline for the design of biofuel production systems.

References

1. Alakangas, E.: European standards for fuel specification and classes of solid biofuels. In: Grammelis, P. (ed.) Solid Biofuels for Energy. Green Energy and Technology. Springer, London (2011)
2. Alakangas E.: Properties of wood fuels used in Finland—BIOSOUTH—project (Technical Research Centre of Finland, VTT Processes, Project report PRO2/P2030/05 (Project C5SU800), Jyvaskyla, 2005)
3. Mott, R.A., Spooner C.E.: The caloric value of carbon in coal. The Dulong relationship. Fuel **19**, 226–231, 242–251 (1940)
4. Lloyd, W.G., Davenport, D.A.: Applying thermodynamics to fossil fuels: heat of combustion fromelemental composition. Chem. Ed. **57**, 56–60 (1980)
5. Bychkov, A.L., Denkin, A.I., Tikhova, V.D., Lomovsky, O.I.: Prediction of higher heating values of plant biomass from ultimate analysis data. J. Therm. Anal. Calorim. **130**, 1399–1405 (2017)
6. Fried A.l, Padouvas E., Rotter H., Varmuza K.: Prediction of heating values of biomass fuel from elemental composition. Anal. Chim. Acta **544**, 191–198 (2005)
7. Acar S., Ayanoglu A.; Determination of higher heating values (HHVs) of biomass fuels. Energy Educ. Sci. Technol. Part A: Energy Sci. Res. **28**(2), 749–758 (2012)
8. Martinez-Allier T., Muradian R.: Handbook of Ecological Economics (Edward Elgar Publ., 2015)
9. Cleveland C.J., Costanza, C.A., Hall R., Kaufmann R.: Energy and the U.S. economy: a biophysical perspective. Science **225**, 890–897 (1984)
10. Murphy D.J., Hall C.A.: Year in review—EROI or energy return on (energy) invested. Ann. N.Y. Acad. Sci. 1185 (2010)
11. Zhang, Y., Colosi, L.M.: Practical ambiguities during calculation of energy ratios and their impacts on lifecycle assessment calculations. Energy Policy **57**, 630–633 (2013)
12. Murphy, D.J., Hall, C.A.S., Dale, M., Cleveland, C.: Order from chaos: A preliminary protocol for determining the EROI of fuels. Sustainability **3**, 1888–1907 (2011)
13. Arodudu, O., Voinov, A., van Duren, I.: Assessing bioenergy potential in rural areas a NEG-EROEI approach. Biomass Bioenerg. **58**, 350–364 (2013)
14. Arodudu, O., Ibrahim, E., Voinov, A., van Duren, I.: Exploring bioenergy potentials of built-up areas based on NEG-EROEI indicators. Ecol. Ind. **47**, 67–79 (2014)
15. Herendeen, R.A.: Net energy analysis: concepts and methods. In: Ayres, R.U., Costanza, R., Goldemberg, J., Ilic, M.D., Jochem, E., Kaufmann, R., Lovins, A.B., Munasinghe, M., Pachauri, R.K., Pardo, C.S., Peterson, P., Schipper, L., Slade, M., Smil, V., Worrell, E., Cleveland, C.J., (eds.) Encyclopedia of Energy. Elsevier, Amsterdam, The Netherlands (2004)
16. Huettner, D.A.: Net energy analysis: an economic assessment. Science **192**, 101–104 (1976)
17. Cleveland, C.J.: Energy quality and energy surplus in the extraction of fossil fuels in the US. Ecol. Econ. **1992**, 139–162 (1992)

18. Brandt, A.R., Dale, M.: A general mathematical framework for calculating systems-scale efficiency of energy extraction and conversion: Energy Return on Investment (EROI) and other energy return ratios. Energies **4**, 1211–1245 (2011). https://doi.org/10.3390/en4081211
19. Bentsen, N.S., Møller, I.M.: Solar energy conserved in biomass: sustainable bioenergy use and reduction of land use change. Renew. Sustain. Energy Rev. **71**, 954–958 (2017)
20. Wasiak A. unpublished work
21. Wasiak A., Orynycz O.: Computer modelling of the influences of subsystems' interactions on energetic efficiency of biofuel production systems. In: ITM Web of Conferences, 01 Jan 2017, vol. 15, p. 04004, ISSN: 2271-2097; E-ISSN: 2271-2097. https://doi.org/10.1051/itmconf/20171504004

Chapter 5
Mathematical Modeling of Energetic Efficiency of Biofuel Production System

5.1 General Description

Life cycle analysis (LCA) is the most frequently used method of modeling, i.e., energetic efficiency of production systems. Except for the energetic characteristics (EROEI), it provides information about the ecological influence of the process being investigated. Since it is based on strictly formalized (normalized) procedures [1, 2], it may serve as a tool for characterization of particular, well-defined situations of real industrial processes. This methodology is, however, not well suited to the analysis of virtual situations that may provide a scientific tool enabling studies on dependencies basing on "what—if" procedures.

This kind of tool is developed, and described in the following parts of the present chapter. The principle of the approach is to build dependencies starting from possibly elementary "ab initio" assumptions, and on choosing the most important factors at the beginning of the analysis, adding other effects in subsequent steps. The idea of partial energetic efficiency indicators, described in the previous chapter, seems to be a very convenient tool for such method of analysis.

The biofuel production system is built of at least two main subsystems: agricultural and industrial, mutually connected by fluxes of materials, energy, and the transportation of various goods (machinery, production means, crop protection means, etc.). Its structure is illustrated in Fig. 5.1.

The typical characteristic of the production system containing agricultural part is the difference in the time scales of operations performed on subsystems. The agricultural activity is seasonal, while the industrial subsystem tends to be yearlong active. The combination of both requires, therefore, some kind of storage between them. Such storage usually also requires inputs of energy that have to be taken into account.

A. Wasiak, *Modeling Energetic Efficiency of Biofuels Production*, Green Energy and Technology, https://doi.org/10.1007/978-3-319-98431-5_5

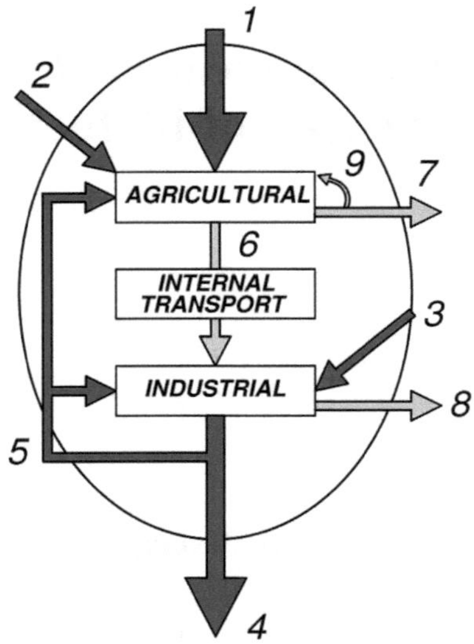

Fig. 5.1 Schematic representation of biofuel production system

5.2 Agricultural Subsystem

Shown in the Chap. 2, Tables 2.12, and 2.14 indicate that the production of "energetic" crop requires several agro-technical operations for tillage, crop protection, and harvesting. Each of those requires specific machinery and time for performing the work. Obviously, some amount of energy has to be spent during each operation. The amount of energy required for a given operation is mainly connected with the fuel consumption, and specific yield of particular machinery.

5.2.1 Partial Energetic Effectiveness of Agricultural Operations

When the shape of the field can be represented by a parallelogram shown in Fig. 5.2, and each operation is realized by driving the machine in parallel to one of the sides, (e.g., C) the area, a, of the elaborated strip of the width, d, equals

$$a = C \times d = \frac{D \times d}{\sin \alpha} \tag{5.1}$$

where C is the length of the side. Obviously, the total area will be covered by, n, strips. The number of strips can be calculated from the dependence as follows:

$$n = \frac{L}{d} \sin \alpha \tag{5.2}$$

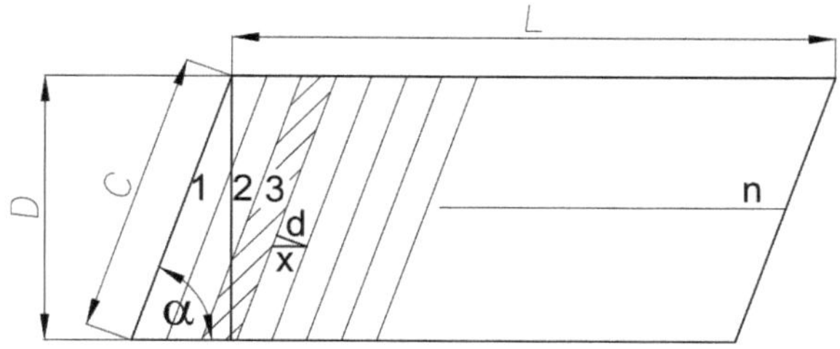

Fig. 5.2 The scheme of the operation on the field

where L is the second side of the parallelogram.

Therefore, the distance driven over the whole field [3, 4] during one operation equals to

$$D_{\text{agr}} = n \times C = \frac{L \sin \alpha}{d} \times \frac{D}{\sin \alpha} = \frac{\text{LD}}{d} \tag{5.3}$$

Since the product LD is equal to the field area, A, the amount of energy, $E_{\text{ex,agr},i}$, that has to be expended through fuel consumption during the particular ith operation performed in the field is equal to

$$E_{\text{ex,agr},i} = W_{\text{fuel}} \times A \times \frac{\omega_i}{d_i} \tag{5.4}$$

and for the set of, m, operations performed on that field, can be expressed as

$$E_{\text{ex,agr}} = W_{\text{fuel}} \times A \times \sum_{i=1}^{m} \left[\frac{\omega_i}{d_i} \right] \tag{5.5}$$

where W_{fuel} is the calorific value of the fuel, and ω_i—is the fuel consumption on the unit of distance in the ith operation (d_i—is the strip width in the ith operation).

The energy derived from particular plantation can be estimated based on the biomass yield (obviously, this part of produced biomass that will be converted into the final form of energy). The amount of that energy can be computed is based on the following equation:

$$E_{\text{bio}} = A \times M_{\text{crop}}\left(c_f, c_w, c_{\text{cp}}\right) \times \gamma \times \sum_{k=1}^{n} \alpha_k \times W_{\text{bio},k} \tag{5.6}$$

According to Eq. (5.6), E_{bio}, depends on plantation area, A, specific yield, M_{crop}, of the mass of biomass corresponding to the unit of plantation area. The M_{crop}, is, in general, a function of several variables like availability of fertilizer, water, and crop protection means, which probably can be expressed in form of actual or average concentrations of corresponding substances in the air or soil. The second term in Eq. (5.5) converts the mass of crop into energy. For this reason, the mass of crop is multiplied by γ—denoting the mass fraction of biofuel in the crop, and in general case, when several types of biofuels are produced—there is a sum of products of low calorific value times the mass fraction, α_k, of kth species of biofuel in the whole amount of biofuels produced from a given crop.

Finally, the partial energetic efficiency for one agricultural operation can be written as follows:

$$\varepsilon_{\text{agr},i} = \frac{M_{\text{crop}}\left(c_f, c_w, c_{cp}\right) \times \gamma \times \sum_{k=1}^{n} \alpha_k \times W_{\text{bio},k}}{W_{\text{fuel}} \times \frac{\omega_i}{d_i}} \tag{5.7}$$

and for the number of, m, operations concerning only agricultural ones:

$$\varepsilon_{\text{agr}} = \frac{M_{\text{crop}}\left(c_f, c_w, c_{cp}\right) \times \gamma \times \sum_{k=1}^{n} \alpha_k \times W_{\text{bio},k}}{W_{\text{fuel}} \times \sum_{i=1}^{m} \frac{\omega_i}{d_i}} \tag{5.8}$$

Equations (5.7) and (5.8) show that in both cases, i.e., an individual agro-technical operation, as in the assembly of such operations the partial energetic effectiveness does not depend upon plantation area. Conditions, at which some deviations from this rule could exist should be studied separately.

5.2.2 *Partial Energetic Effectiveness of Internal Transport*

During agro-technical operations carried out on the fields, frequent transportation of goods outside of the field must occur. It concerns driving of machinery to the field and backward, transporting various goods like fertilizers, crop protection means, etc., as well as transporting crops harvested from the fields. In the large plantations needs for multiple courses of transportation, means may be necessary.

Multiple courses might be needed because a given operation carried out on some field may last longer than normal allowable working time, and several drives toward workplace and back have to occur. It may also be caused by large amounts of goods that have to be transported. When the load capacity of available transportation means is exceeded, multiple courses are indispensable. Figure 5.3 illustrates the situation described in the former case. On the first day the machine travels the distance, c, from the base to the field, and elaborates some fraction of that field starting at some initial point, and after the working hours passed, returns to the base from the point different than the starting one. The returning distance is then $c + a_1/J$. The, J, is a number of days that is needed for the elaboration of the whole field, and a_1/J is the fraction of the distance between the beginning and the end of the field.

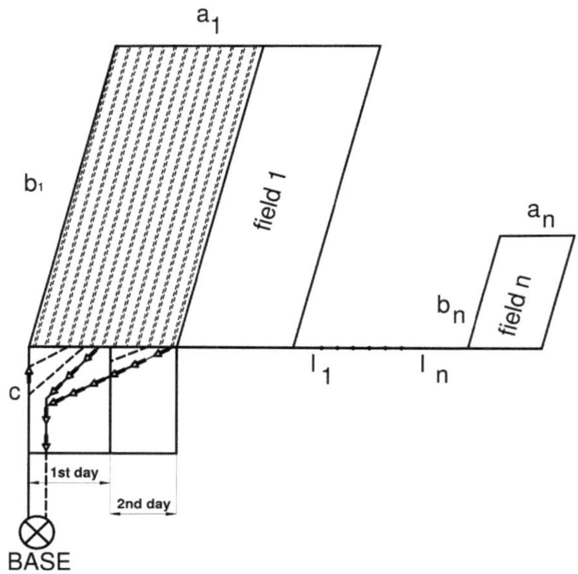

Fig. 5.3 Schematic representation of the distance driven on the field and outside of the field in the case when each of the fields in the plantation requires several days for a given operation

On the next day drives from the base to the point where yesterday the work was stopped. A similar situation can be repeated several times. In the case, when plantation contains several fields, at some instant of time the machine drives from previous to the next field adding the distance, l_1, between the fields to the previous scheme of traveling. In such a case, the distance driven during one (long-lasting) agro-technical operation can be expressed [5] as

$$D_{\text{out}} = 2c\sum_{n=1}^{N} J_n + 2\sum_{n=1}^{N-1}\left[(a_n + l_n)\sum_{n+1}^{N} J_n\right] + D_{\max}\sum_{n=1}^{N} J_n^2 \frac{d_n}{b_n} \tag{5.9}$$

The number of days needed is determined by the distance driven on the field, D_{agr}, given by Eq. (5.3), and the average speed of field operations. The maximum distance, $D_{\max}$, driven on the field is, in turn, given by the assumed maximum daily working time, and the average speed of the equipment. The number of days, J_n, required for the carrying out of operations on the nth field is given as

$$J_n\left[\text{days}\right] = \frac{D_{\text{agr},n}\,[\text{km}]}{v\,[\text{km/h}] \times t_{\max}\left[h/\text{day}\right]} \tag{5.10}$$

The partial energetic effectiveness of internal transport operations for the case of the field machining is, therefore, given as

$$\varepsilon_{\text{tr}} = \frac{M_{\text{crop}}\left(c_f, c_w, c_{\text{cp}}\right) \times \gamma \times \sum_{k=1}^{n} \alpha_k \times W_{\text{bio},k}}{\{2c \sum_{n=1}^{N} J_n + 2 \sum_{n=1}^{N-1} [(a_n + l_n) \sum_{n+1}^{N} J_n] + D\text{max} \sum_{n=1}^{N} J_n^2 \frac{d_n}{b_n}\} \times W_{\text{fuel}}} \tag{5.11}$$

As seen from Eq. (5.11), energetic efficiency of internal transport operations is dependent on plantation structure—field sizes, number of fields and distances between them, on the type of agricultural equipment. The estimation of the importance of this contribution can be obtained more precisely on the basis of numerical computations.

Results of computations may, of course, be validated with the empirical data from direct measurements performed in the real farm.

5.2.3 Numerical Computations

The model described in previous parts of this chapter enables numerical computations with the use of the data representing the real situation or modeling of the behavior of the production system under changes of chosen variables. The latter method gives the opportunity to recognize the character of the dependencies of energetic efficiency being function of variables of interest. Figure 5.4 shows the flowchart of the computer program used for computations [6]. The program uses the data describing plantation topology and dimensions (shown in Fig. 5.5), and characteristics of the equipment used for tillage of the fields according to equations Eqs. (5.1–5.11). After the data input at first, the daily operating distance is computed, and after, in two nested loops, values of distances driven in the fields, D_{agr}, and distances driven outside the fields, D_{out}, are evaluated. Further, the values of energy contained in the form of biofuel, E_{bio}, as well as energies consumed during operation on the fields, E_{agr}, and outside the fields, E_{out}, are computed. Finally, the energetic efficiencies ε_{agr}, ε_{out}, and the aggregated one, $\varepsilon_{\text{agr+out}}$, are evaluated.

Results of computations were published in several papers [4–7]. The structure of plantation shown in Fig. 5.6, is composed of five fields of equal sizes with dimensions $b = 0.5$ km. The variable a, the distance, l, and the distance from the base to the first field, c, were also variable. Several values of the width of the operation strip, d, were also assumed.

The low caloric value of the fuel used for running machines was taken as 36 MJ/dm^3, and for biofuel 34.6 MJ/dm^3. Fuel consumption by agricultural machines was assumed 0.3 dm^3/km. Machine speed was estimated as 6 km/h, and maximum daily allowable working time was accepted as 10 h. The resulting daily working distance driven on the field was equal to 60 km.

An example of computed distances driven on the field are given in Table 5.1 where values, E_{bio}, distances driven on the fields of different areas, D_{agr}, energy consumed

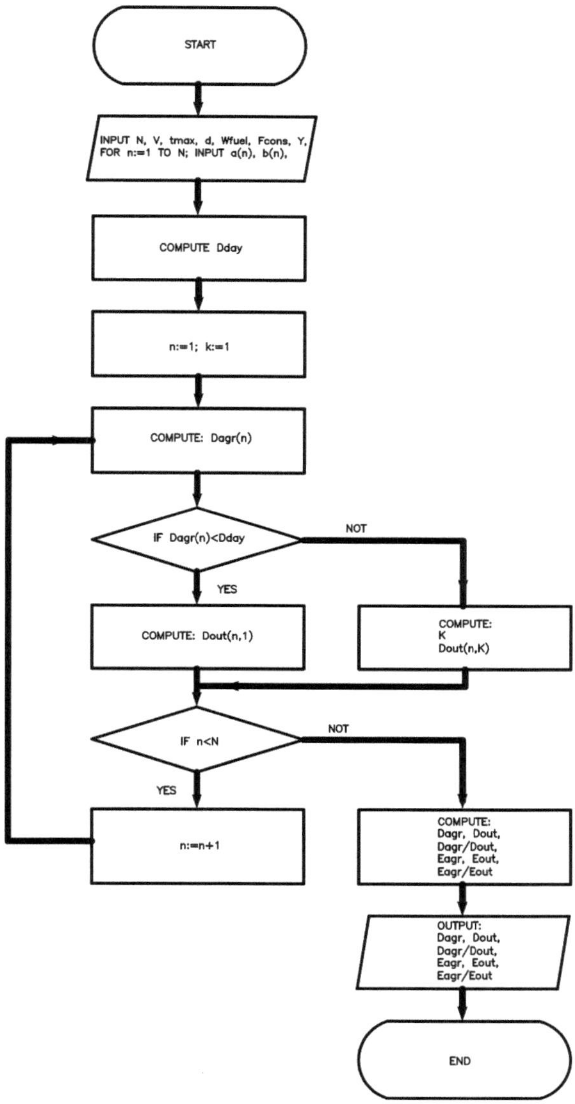

Fig. 5.4 Flow chart of the simulation program [reproduced from Ref. [6] (All articles published in Applied Computer Science are open access and distributed under the terms of the Creative Commons Attribution 4.0 International License.)]

during driving machines on the field, E_{agr}, for two values of the field area, as well as corresponding values of partial energetic efficiency, ε_{agr}, are listed.

Further, the distances driven outside of the fields are given in Table 5.2. It is seen that the distances driven on the fields (given in Table 5.1.) are substantially bigger than those, driven outside. It is also clear, that the wider is the operational strip, the shorter are both the distances.

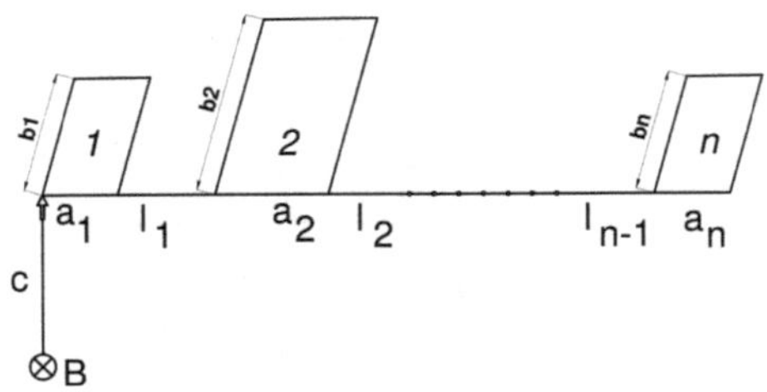

Fig. 5.5 The assumed topological structure of plantation chosen for computer simulations

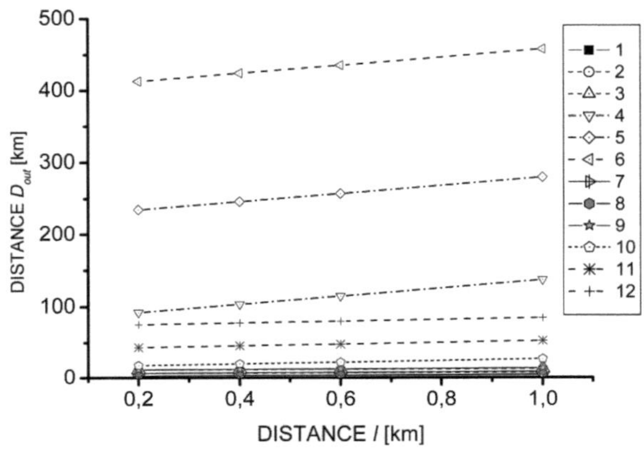

Fig. 5.6 The dependence of the distance driven outside fields upon other dimensional characteristics of the plantation

Table 5.1 Partial energetic efficiency of one agricultural operation carried out on a rapeseed plantation consisting of 5 equal fields each of area A

	A [km^2] =	0.01	0.25
d [m]	E_{bio} (5 fields) [GJ] =	259.4	6485.6
0.5	D_{agr} [km] =	100	2500
4	D_{agr} [km] =	12.5	312.5
0.5	E_{agr} (5 fields) [MJ] =	1080	27000
4	E_{agr} (5 fields) [MJ] =	135	3375
0.5	ε_{agr} (5 fields) =	240.3	240.3
4	ε_{agr} (5 fields) =	1921.7	1921.7

The distances driven outside of the field are especially high for the large fields, elaborated with the small width of the operating machine. These distances also strongly increase with an increase of the distances, c, and, l. This behavior is mainly caused by multiple returns after finishing the working day. Obviously, it indicates the incorrect choice of the equipment.

Table 5.2 The distance driven outside of the fields in relations to the field area, the distances between fields, and the width of operation strip

		A = 0.01	A = 0.01	A = 0.01	A = 0.25	A = 0.25	A = 0.25
d [m]	l [km]	c = 1	c = 5	c = 10	c = 1	c = 5	c = 10
0.5	0.2	2.4	6.4	11.4	92.18	234.58	412.58
0.5	0.4	2.8	6.8	11.8	103.26	245.66	423.66
0.5	0.6	3.2	7.2	12.2	114.34	256.74	434.74
0.5	1	4	8	13	136.5	278.9	456.9
4	0.2	2.6	6.6	11.6	17.72	43.32	75.32
4	0.4	3.2	7.2	12.2	20.04	45.64	77.64
4	0.6	3.8	7.8	12.8	22.36	47.96	79.96
4	1	5	9	14	27	52.6	84.6

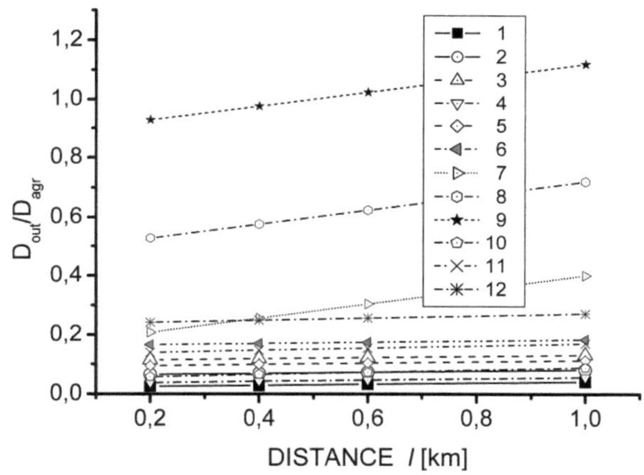

Fig. 5.7 The ratio of the distances D_{out}/D_{agr} driven outside and in the fields as a function of the distance between fields

Figure 5.6 shows the dependence of the distance, D_{out}, driven outside the fields, upon the other parameters of the plantation structure, i.e., fields areas, distances between fields, the distance between the base and the first field. The small values of this distance are observed for small fields elaborated with narrow and wide operation strip. Larger values are seen in large fields operated with the wide strip, and finally much larger values—when large fields are operated with a narrow strip. The large values of, D_{out}, are mainly due to multiple returns from the field to the base after approaching allowable daily work time.

The ratio of the distance driven outside the fields to this, driven on the fields are presented in Fig. 5.7. Majority of the dependences show the weak effect of the fields dimensions and rather small values of this ratio. For the few curves on the plot showing slightly higher values of the ratio, the highest values of the distance, *c*, seems to be responsible.

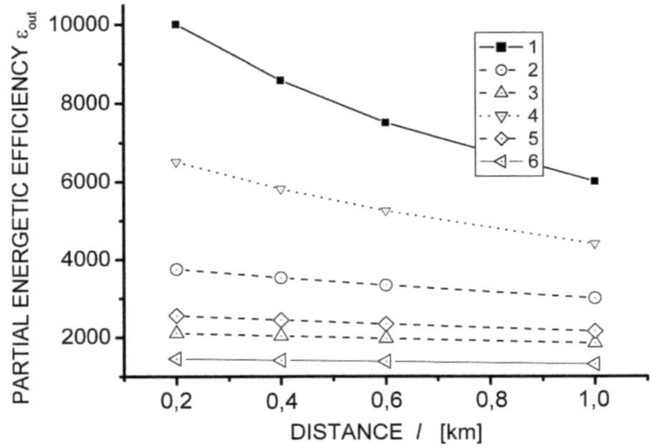

Fig. 5.8 The partial energetic efficiency, ε_{out}, for one internal transport operation computed for the rapeseed plantation model shown in Fig. 5.5. The plot contains results for narrow operation strip

Numbers seen on the legend of both pictures denote the following characteristics of work field:

1 - c = 1 km A = 0.01 km^2 d = 0.5 m,
2 - c = 5 km A = 0.01 km^2 d = 0.5 m,
3 - c = 10 km A = 0.01 km^2 d = 0.5 m,
4 - c = 1 km A = 0.25 km^2 d = 0.5 m,
5 - c = 5 km A = 0.25 km^2 d = 0.5 m,
6 - c = 10 km A = 0.25 km^2 d = 0.5 m,
7- c = 1 km A = 0.01 km^2 d = 4.0 m,
8 - c = 5 km A = 0.01 km^2 d = 4.0 m,
9 - c = 10 km A = 0.01 km^{2}d = 4.0 m,
10 - c = 1 km A = 0.25 km^2 d = 4.0 m,
11 - c = 5 km A = 0.25 km^2 d = 4.0 m,
12 - c = 10 km A = 0.25 km^2 d = 4.0 m.

Most of the values of the ratio mentioned above, are rather small, they are located within limits $0.02 < D_{out}/D_{agr,} < 1.7$. Those limits are rather results of specific values of parameters taken into consideration, and cannot be considered as any universal constants.

The values of energy consumed during an operation on the field, $E_{agr,}$ and during transportation outside the fields, E_{out}, are correspondingly proportional to the distances driven in both cases. Those values were computed, and used to compute partial energetic efficiency for both: agro-technical operations, ε_{agr}, and the internal transport courses ε_{out}.

The dependence of the partial energetic efficiency, ε_{out}, for the small operational width, *d,* analyzed in the model calculations is shown in Fig. 5.8.

In this case, the energetic efficiency reaches high values that slightly decrease with an increase of the distance, *c,* i.e., the distance between the base and the first operated field. It also decreases with an increase of the area of the field, and with an increase of distances between fields. It has to be pointed out that presented values

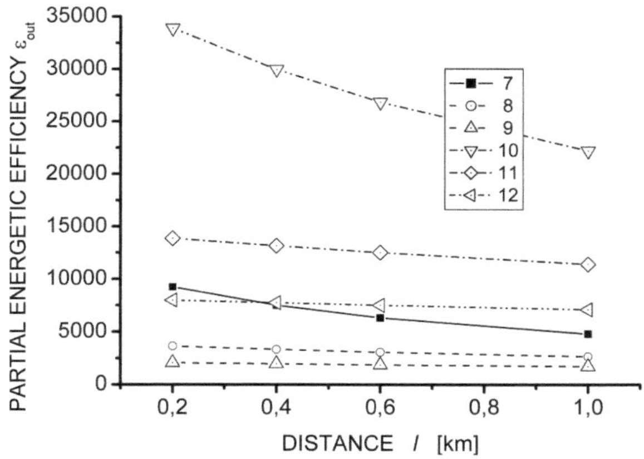

Fig. 5.9 The partial energetic efficiency, ε_{out}, for one operation with the strip width 4 m on the plantation consisting of five fields. The results for 4 m operation strip concern only one agro-technical operation carried out on all five fields. The increase in the number of operations, that is usually the normal situation, decreases global energetic efficiency, accordingly to the explanation given in Chap. 4 (Eq. 4.13).

Figure 5.9 shows the computed energetic effectiveness of internal transport for the model of rapeseed plantation elaborated with the machine working 4 m wide operation strip. Again results concern one agro-technical operation carried out in five fields. As it was mentioned earlier, the internal transport operations are associated with tillage operations, and, therefore, depend on plantation topological structure, as well as sizes of the fields, and also on technical characteristics of machinery applied for the field work. The highest efficiency is shown by the large plantation, in which distances between the fields, and the distance between the base and the first field are the smallest. The values of, ε_{out}, reached in the case of a wider operation strip are much higher than those with the narrow strip.

Figures 5.10 and 5.11, in turn, show the results of the combination of the efficiency of agricultural operations, $\varepsilon_{\text{agr,}}$ with the efficiency of the internal transport, ε_{out}.

The energetic effectiveness, in this case, is evidently lower than the one for internal transport alone. The behavior is rather typical, i.e., efficiency decreases with an increase of the distances between fields, and the distance between the base and the first field.

Since in the normal practice several agro-technical operations are carried out during production year one has to consider a substantial decrease of energetic efficiency of the whole production system. Assuming that all operations require the same amount of energy (which fortunately is not the case), the resulting energetic effectiveness of agricultural part of the production system will decrease as shown in Table 5.3.

Not only energy consumption is a factor limiting the choices for agro-technical machinery. In the cases of the narrow working strip, the long driving distances require a long time to complete the operation. For large fields, this time can be quite long. Having in mind that each agricultural operation must be carried out in appropriate

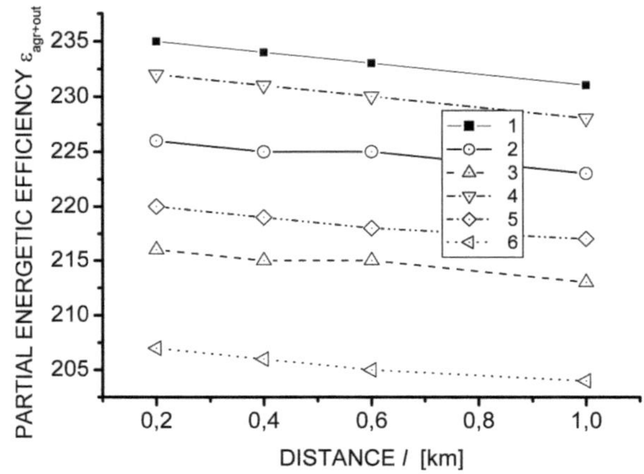

Fig. 5.10 The partial energetic efficiency for the case of one agro-technical operation with the narrow operation strip, combined with corresponding internal transport carried out on the rapeseed plantation consisting of five fields of the same size

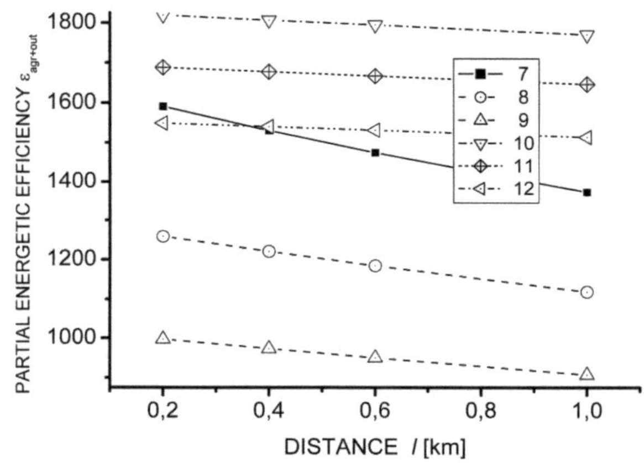

Fig. 5.11 The partial energetic efficiency for the case of one agro-technical operation with the narrow operation strip, combined with corresponding internal transport carried out on the rapeseed plantation consisting of five fields of the same size

season corresponding to biological requirements of a particular plant, as well as the all necessary tillage operations have to fit to the length of the specific season of the year some solutions must be excluded because they would not fit to biological performance "window".

Table 5.4 shows the duration of operations depending on the width of the operation strip and the area of plantation. The figures given in boldface show the durations of the operation that have to be avoided.

It has to be noted, however, that in the mutual interactions of agro-technical and internal transport operations the agro-technical ones are predominant with respect of the energetic effectiveness of the production system (or at least its agricultural part). Consequently, in the attempts of optimization the agro-technical operations: tillage, etc., has to be considered as a first priority. Such optimization may be achieved by reduction of the number of operations (e.g., some kind of hybridization), and also

Table 5.3 A decrease of energetic effectiveness with increasing number of identical operations

		No	A = 0.01			A = 0.25		
d [m]	l [km]	Operations	c = 1	c = 5	c = 10	c = 1	c = 5	c = 10
0.5	0.4	**1**	**234**	**225**	**215**	**231**	**219**	**206**
0.5	0.4	2	117	113	108	116	110	103
0.5	0.4	3	78	75	72	77	73	69
0.5	0.4	4	59	57	54	58	55	52
0.5	0.4	5	47	45	43	47	44	42
4	0.4	**1**	**1530**	**1220**	**973**	**1806**	**1677**	**1540**
4	04	2	765	610	487	903	839	770
4	0.4	3	510	407	325	602	559	514
4	0.4	4	383	305	244	452	420	385
4	0.4	5	306	244	195	362	336	308

Table 5.4 Number of days needed to perform the agricultural operation, as a function of plantation size, and operation width of the equipment

A[km^2]	Number of days J				
	d = 0.5	d = 1	d = 1.5	d = 2	d = 2.5
1.25	**45**	25	15	15	10
2.25	**75**	**40**	25	20	15
2.75	**95**	**50**	35	25	20
3.25	**110**	**55**	**40**	30	25
3.75	**125**	**65**	**45**	35	25

Source Wasiak [7] Proceedia Engineering Vol. 182 (2017) pp. 739–746 (Open access)

by the appropriate choice of operating machines. Computer modeling may happen to be a good tool for such tasks.

5.2.3.1 External Transport

External transport from the viewpoint of the agricultural producer of energetic crop-s—is mainly the transport of crops between plantation and industrial facility that converts biomass onto biofuel of any type. The other transportation needs rather do not reach such a big volume of the load as it is in the case of crops. The amount of trucks needed to transport rapeseed grain from plantations of various sizes is given in Table 5.5. Approximate values of specific fuel consumption are also given thereof.

Based on the data from Table 5.5 and the values of the calorific value of the fuel the partial energetic efficiency, $\varepsilon_{ext,tr,}$ for several examples of plantations were computed. The selected results are given in Table 5.6. The energetic efficiency is practically independent of plantation area, and the load capacity of a truck. It only decreases with an increase in transportation distance. The behavior of partial energetic efficiency

Table 5.5 Number of trucks of various load capacities needed for transport of rapeseed grain from fields of different area

			Number		
Fields [ha]	Grain [Mg]	Grain [m3]	3 ton truck's	10 ton trucks	30 ton truck
Fuel consumption [l/km] =			0.1	0.2	0.3
50	150	229	23	11.5	7.7
100	300	458	45.9	23	15.3
200	600	916	91.7	45.9	30.6

Source prepared on the basis of [8]

Table 5.6 Partial energetic effectiveness of eternal transport of rapeseed grain for chosen areas of plantations

		External distance 50 km			External distance 100 km		
Field [ha]	Grain [Mg]	3 ton trucks	10 ton trucks	30 ton truck	3 ton trucks	10 ton trucks	30 ton trucks
50	150	418.2	418.2	416.4	209.1	209.1	208.2
100	300	419.1	418.2	419.1	209.6	209.1	209.6
200	600	419.6	419.1	419.1	209.8	209.6	209.6

observed in this case requires further investigations, since the present results might be a consequence of the fortunate choice of sizes and fuel consumption of the trucks discussed.

Again, taking into account that the values established in Table 5.6 should be correspondingly introduced to the aggregate values of agricultural energetic efficiency, one might expect a further decrease of the global efficiency of a production system.

References

1. Curran M.A.: Life cycle assessment: principles and practice; Scientific Applications International Corporation (SAIC); EPA Contract No. 68-C02-067 Work Assignment 3–15 (2006)
2. Grant T., Bengtsson J., Logie J., Cowie A., Henke J. M.: Life cycle assessment (LCA) of bioenergy products and projects. Australian Renewable Energy Agency (2016, October)
3. Wasiak A., Orynycz O.: Formulation of a model for energetic efficiency of agricultural subsystem of biofuel production. IEEE International Energy Conference: ENERGYCON'2014, Croatia, pp. 1333–1337 (2014)
4. Wasiak, A., Orynycz, O.: The effects of energy contributions into subsidiary processes on energetic efficiency of biomass plantation supplying biofuel production system. Agric. Agric. Sci. Procedia **7**, 292–300 (2015)
5. Wasiak, A., Orynycz, O.: Energy efficiency of a biofuel production system. Manage. Prod. Eng. Rev. **1**(2), 10–20 (2010)

6. Wasiak, A., Orynycz, O.: Effects of field's topology on energetic efficiency of rapeseed plantation for biofuel production. Appl. Comput. Sci. **11**(1), 83–98 (2015)
7. Wasiak, A.L.: Effect of biofuel production on sustainability of agriculture. Procedia Eng. **182**, 739–746 (2017)
8. Orynycz, O., Wasiak, A.: The effect of external transport on energetic efficiency of biodiesel production. J. Ecol. Eng. **18**(2), 57–62 (2017)

Chapter 6
The Effects of Embodied Energy

6.1 Main Principles

According to [1], the embodied energy, called also "emergy", is the amount of available energy (exergy) that have been used during the whole cycle of production of some object, including the energy consumed for extraction, transportation, etc., of resources needed for production. Some works [2] suggest adding also the energy introduced with the spare parts and operations needed for repairs.

Exergy, in turn, is defined as this part of the energy that can be converted into work. Based on considerations given in [3] for the case of thermal exergy it can be written as follows:

$$\Delta B_{th} = \Delta I - T_{env}\Delta S \quad (6.1)$$

where ΔB_{th}—is the change of exergy, ΔI and ΔS—corresponding changes of enthalpy and entropy, and T_{env}—the temperature of the natural environment.

One can recognize that it is equivalent to Gibbs free energy (free enthalpy) with conditions of natural environment accepted as a reference state.

Practical calculations are usually based [2] on energy consumption in industrial sectors involved in the production of raw materials, as well as the corresponding technical product (machine).

Many substances like materials and other production means (in agriculture: fertilizers, crop protection means) are completely used up in a very short time. In such a case, the computation of their contribution to energy efficiency is relatively simple. The example of calculation of embodied energy contribution due to the application of chemical fertilizers and herbicide is shown in Table 6.1. The data from Ref. [4] concern plantation of Miscanthus. Since the amount of energy per annum, that can be obtained from the crop is 322.9 GJ/ha/a, and the sum of the contributions listed in

A. Wasiak, *Modeling Energetic Efficiency of Biofuels Production*, Green Energy and Technology, https://doi.org/10.1007/978-3-319-98431-5_6

Table 6.1 Embodied energy contained in some chemicals used

Material	Energy content [MJ/kg]	Dosage [kg/ha]	Energy consumption [MJ/ha]
Herbicide (Glyphosate)	454	20	9080
Propagation means	0.0862	800	68.96
Nitrogen fertilizer	78.1	50	3905
Phosphorus fertilizer	17.4	21	365.4
Potassium fertilizer	13.7	45	616.5

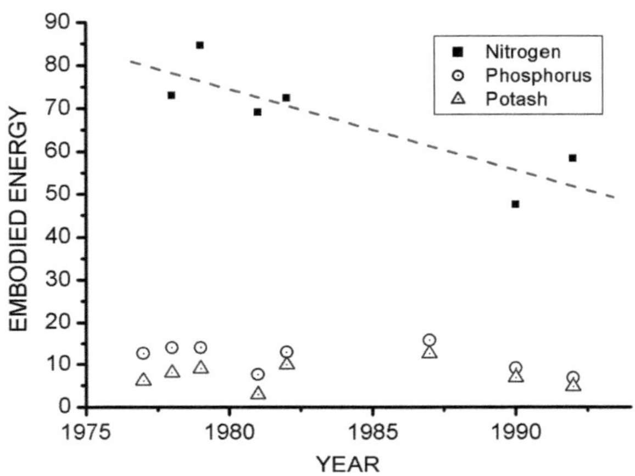

Fig. 6.1 Embodied energy contained in some fertilizers as a function of production year

Table 6.1 is 14.03 GJ/ha/a the partial energetic efficiency resulting of this contribution equals 20.0. Values of embodied energy content can be obtained from various publications, e.g., [5]. It has to be taken into account that data obtained from different sources might differ. This is because they are affected by the local level of technology (in a given country), and may also vary in time [5]. An example of such changes occurring during several production years is shown in Fig. 6.1.

For the case of long-lasting agricultural machinery, the embodied energy is usually estimated for the whole life cycle of the machine or foreseen the number of hectares operated. Since the machine is usually composed of several materials, and some energy has to be used for assembling, calculation of embodied energy content appears to be slightly more complicated. The contributions of the individual component have to be computed by multiplication of a value obtained for individual material times the mass fraction of this material in the product.

The examples of embodied energy density [5] of several construction materials are given in Table 6.2.

Table 6.2 Embodied energy densities for several construction materials

Material	MJ/kg	GJ/m^3
Aluminum	227	515.7
Aluminum recycled	8.1	21.87
Brass	62	519.56
Copper	70.6	631.16
Glass	15.9	37.55
PVC	70	93.62
Steel	32	251.2
Steel recycled	8.9	37.21
Zinc	51	371.28

Source Own elaboration based on the data from [5]

Such data, together with the fraction of the material in the machine, and obviously the total mass, and expected time of useful life of the machine, are used for calculation of the global embodied energy content of that machine. Some examples are given in Table 6.3 [6].

The data contained in Table 6.3 present the expected lifetime of each piece of equipment, its mass, embodied energy intensity corresponding to materials, manufacturing (assembling), and finally the part needed for repairs and maintenance. The latter is the most difficult to estimate. There is not enough data accumulated that concern this problem. It is, however, more frequently discussed in economic categories. Several papers published, address the cost of repair and maintenance. For example, the paper [7] relates accumulated repair costs, Kn, to the list price, Cm, of the machine defining the repair cost coefficient, r, as equal to

$$r = \mathrm{Kn}/\mathrm{Cm} \tag{6.2}$$

And further concludes that unit repair cost can be written as

$$\mathrm{Kjn} = \mathrm{Cm} \times r/W_T \tag{6.3}$$

where
Kjn—repair unit cost [monetary units/h]
W_T—estimated lifetime [h].

The paper estimates the value of coefficient, r, approximately about five, however, the data for various countries are not available.

In several papers [8], the time dependence of repair costs is given as empirical equations. According to the paper [9], the best fit is achieved when the following power formula is used:

$$Y(x) = x^b \tag{6.4}$$

Table 6.3 Embodied energy content of chosen agro-technical machines

Type of machine	Energy input						
	Useful life [h]	Useful life [years]	Power [kW]	Weight [kg]	Raw materials [MJ/kg]	Manufacturing [MJ/kg]	Repairs and maintenance [MJ/kg]
Centrifugal fertilizer spreader	2000	6	1.84	10	62.8	7.4	26
Chainsaw	2500	12	0	700	62.8	8.6	30
Cultivator	2500	12	0	700	62.8	8.6	30
Disc harrow	2000	6	1.84	10	62.8	7.4	26
Handheld trimmer	2500	15	0	700	62.8	8.6	30
Roller	1500	10	0	400	62.8	8.6	30
Sprayer tank (1000 l)	1500	10	0	100	62.8	8.6	30
Sprayer tank (100–400 l)	2000	15	0	500	62.8	8.6	30
Tine harrow	12,000	10	41	2300	49.4	14.70	49
Tractor: 2 RM, 41 kW	12,000	10	50	3400	49.4	14.7	49
Tractor: 2 RM, 50 kW	16,000	12	100	5500	49.4	14.7	49
Tractor: 44, 100 kW	16,000	12	50	3900	49.4	14.7	49
Tractor: 44, 50 kW	16,000	12	75	4700	49.4	14.7	49
Tractor: 44, 75 kW	5000	15	0	300	62.8	8.6	30
Trailer (0,7–1 t)	5000	15	0	1600	62.8	8.6	30
Trailer (5–6 t)	5000	15	0	1600	62.8	8.6	30

Source Own elaboration on the basis of the work [6]

which expresses the function increasing in time.

The repair cost expressed in monetary units may be expected to be proportional to the embodied energy associated with the repair and maintenance processes. The proportionality coefficient may, however, be different for various machines as well as may vary with the local conditions.

Assuming that time dependence of embodied energy may be approximated by a continuous function, the following formula can be postulated to express the average intensity of embodied energy:

$$\langle E_{\text{emb}} \rangle = E_{\text{material}} + E_{\text{manufacturing}} + \int_0^{t_{\text{life}}} E_{\text{repair}}(\tau)\mathrm{d}\tau / t_{\text{life}} \tag{6.5}$$

where E_{repair} is a function of time, and t_{life}—is the expected lifetime of the machine.

or

$$\langle E_{\text{emb}} \rangle = E_{\text{material}} + E_{\text{manufacturing}} + \int_0^{\text{AREA}} E_{\text{repair}}(\text{area})\mathrm{d}(\text{area}) / \text{AREA} \tag{6.5}$$

where (area) is the surface area of the elaborated field, and *AREA* is the total area of the fields elaborated during the lifetime of the equipment.

Since the integration limits exceed the actual age of the machine at the instant of time when the calculated value of embodied energy $\langle E_{\text{emb}} \rangle$ is practically applied, the function $E_{\text{repair}}(t)$ or $E_{\text{repair}}(\text{area})$ has to be evaluated on the basis of a large number of historical data enabling eventual extrapolation. Undoubtedly, this topic requires further research.

References

1. Odum, T.H.: Environment, power and society for twenty-first century; the hierarchy of energy. Columbia University Press (2007)
2. Borjesson, P.I.I.: Energy analysis of biomass production and transportation. Biomass Bioenerg **11**(4), 305318 (1996)
3. Szargut, J.: Termodynamika (Thermodynamics) PWN Warszawa (2013)
4. Sopegno, A., Rodias, E., Bochtis, D., Busato, P., Berruto, R., Boero, V., Sørensen, C.: Model for Energy Analysis of Miscanthus Production and Transportation. Energies **9**(39), 2 (2016). https://doi.org/10.3390/en9060392
5. Tables of Embodied Energy Coefficients https://www.victoria.ac.nz/architecture/centres/cbpr/resources/pdfs/ee-coefficients.pdf
6. Guzmán, G.I., Alonso, A.M.: A comparison of energy use in conventional and organic olive oil production in Spain. Agr. Syst. (2008). https://doi.org/10.1016/j.agsy.2008.06.004
7. Lorencowicz, E., Uziak, J.: Repair cost of tractors and agricultural machines in family farms. Agric. Agric. Sci. Procedia **7**, 152–157 (2015)
8. Abdelmotaleb, I. A.: Repair and maintenance cost analysis of tractors and combines. Retrospective Theses and Dissertations. 9260. http://lib.dr.iastate.edu/rtd/9260 (1989)
9. Gautam A.K., Shrivastava A.K.: Development of mathematical model for repair and maintenance of some of the farm tractors of JNKVV, Jabalpur, India. Adv. Crop Sci. Tech. 5, 288. https://doi.org/10.4172/2329-8863.1000288 (2017)

Chapter 7
Sustainability of Agriculture

7.1 Consequences of Biofuel Production

The idea of sustainable development, introduced by Brundtland Report [1], seems to be an important motivation for the search of long-lasting accessibility of natural resources, and maintaining of the quality of the natural environment, and simultaneously achieving the stable economic growth. Those achievements should be accompanied with good quality of life of both: the single man as well as the whole societies. All those factors require, however, the growth of industrial, and agricultural production, and require the growth of supply of energy. Such growth, in turn, is considered to be the cause of depletion of natural resources, and degradation of natural environment. The idea of sustainable development in recent years was complemented or even strengthened by the ideas of green economy and green growth [2]. The trend, being more science-oriented, strongly emphasize the role of energy in the creation of economic development, and social wellbeing. Energy harvesting, in turn, causes depletion of natural resources, and contributes to the degradation of natural environment. In spite of some opinions about the necessity to reduce energy consumption [3], it seems, that contemporary standard of life of the world's population cannot be maintained, and particularly improved (what is desirable in many regions) without further economic growth, which is dependent upon the availability of energy. Taking into account that actual energy supply is mainly originated from fossil resources, like natural gas, petroleum, or coal, one has to keep in mind threats connected with those resources. Two main factors should be mentioned: the first, s.c. "oil peak", denotes the maximum of the rate of resource logging. After the maximum is reached, the yield of the fossil resource will decrease. The existence of such maximum seems to be unquestionable, in spite that some controversies occur concerning temporal perspective of its appearance. The emission of carbon dioxide, that inseparably is connected to the use of fossil resources of energy, is the other factor. The burning of fossil fuels causes an increase in carbon dioxide concentration in both: the atmosphere, and ocean waters. In both cases, an increase of carbon dioxide

A. Wasiak, *Modeling Energetic Efficiency of Biofuels Production*, Green Energy and Technology, https://doi.org/10.1007/978-3-319-98431-5_7

concentration is suspected of bringing dangerous consequences to the environment. The search for technical solutions, which would help to slow down the depletion of natural resources and to reduce the negative impact on environment become an important task for contemporary engineers.

The shift from fossil fuels to alternative fuels is recently indicated as a remedy toward threats mentioned above. The biofuels might play a substantial role in this replacement. Seeds, fruits, as well as other parts of plants [4] are used as resources for biofuel production [5–7] that may contribute to the mitigation of both threats, especially to the appeasement of carbon dioxide emission. Some papers, however, suggest caution with respect to biomass production potential.

7.2 Measures of Sustainability

Understanding of the terms "sustainability" and "sustainable development" may be different, depending on context. Both are sometimes used as synonyms. The meaning of them seems to be distinguished. The term "sustainability" may be understood as expressing a static situation—corresponding to achieved saturation with some respect. In contrast, the term "sustainable development", suggest dynamic situation, during which some characteristics (technological, or economic) are permanently growing, i.e., experience at least a linear increase with time.

Measures indicating the level of sustainability achieved seem to be inadequate, and rather difficult to define. In spite of that simultaneous progress should be achieved in many areas of very different nature, and also some local needs require specific measures. Consequently, quite a large number of indices have been defined [8] giving insight into particular situations, but all those indicators do not form any consequent coordinate system, in which some metric would be defined. Therefore, the progress shown by one of the indexes may simultaneously cause a negative effect in other indices. The system of coordinates that may be associated with the present description of sustainable development may be illustrated by Fig. 7.1. Such a system may correctly describe situation occurring along any one of the axes. The use of such a system requires a correct choice of the most important variables, and also distinguishing between variables denoting causes from those showing the results (or goals).

It is seen that important variables, shown in the Figure, have neither common origin nor common units. It seems, therefore, reasonable to focus on a specific variable that can be quantitatively estimated.

As pointed out in [8] only a few indicators have gained wide acceptance: the first is "ecological footprint", which compares the demand for consumption of various goods on a given area to the biological productivity, the other are emergy performance indices, based on the idea of embodied energy (emergy) [9]. The third is environmental sustainability index that is aggregated index including 21 indicators and 76 variables. It has to be mentioned that the mathematically sophisticated economic model of sustainable growth has been proposed [10].

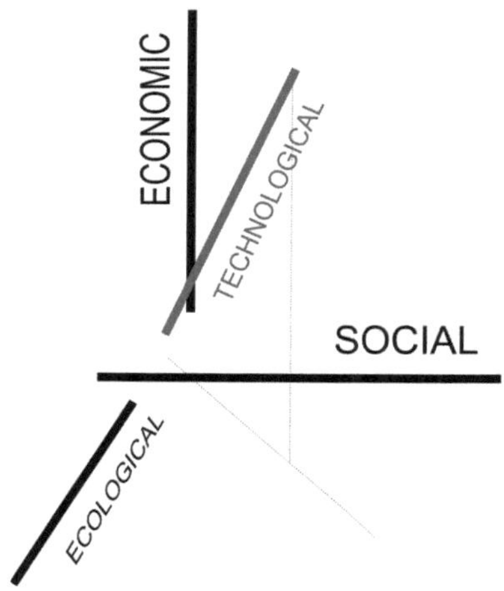

Fig. 7.1 Coordinate system describing the idea of sustainable development

7.3 Effects of Biofuel Production on Sustainability

Analyzing which factors being essential, from the viewpoint of biofuel production, one might come to the conclusion that, at least one of the factors is emission, and subsequent sequestration of carbon dioxide.

The biofuel production, and its subsequent use can be considered as time-dependent processes occurring under conditions of limited resources. This condition is typical for the majority of natural processes.

The progress of such processes can usually be represented by s.c. the sigmoidal curve that is given by the following equation:

$$x(t) = a\left[1 - \exp\left(-kt^n\right)\right] \tag{7.1}$$

in which $x(t)$ represents the amount (fraction) of a component that is formed due to the process, k—is rate constant, and a and n are constant coefficients. The rate constant, k, determines another important characteristic of the process, namely its "halftime"—the time when the fraction of transformed component reaches the value ½. The halftime is given as

$$t_{1/2} = \left(\frac{\ln 2}{k}\right)^{1/n} \tag{7.2}$$

The rate of the process is also a function of time, which can be expressed as

$$\frac{\mathrm{d}x(t)}{\mathrm{d}t} = a\left[-\exp\left(-kt^n\right)\right]nkt^{n-1} \tag{7.3}$$

It is easy to recognize that exponential term in Eq. (7.3) can be replaced by $[1 - x(t)]$, as shown in Eq. (7.4).

$$\frac{dx(t)}{dt} = a[1 - x(t)]nkt^{n-1} \tag{7.4}$$

The rate of the process, as given by Eqs. (7.3) and (7.4) reaches a maximum at $t_{1/2}$ tending to zero below and above the maximum. It is also as evident that it is proportional to the actual fraction of remaining substrate $[1 - x(t)]$. Consequently, the rate of the transition reaches zero when the substrate is exhausted. Such process, from the definition, cannot be considered as a sustainable one.[1]

In the paper [11], the preliminary considerations of the effect of substrate recirculation were introduced. The extended version of the idea, mentioned, is presented below.

The processes occurring in the biofuel production, and their use is associated with the interaction with atmosphere consisting in absorption and emission of carbon dioxide. During biomass production, carbon dioxide is absorbed from the atmosphere, and fixed in the biomass. The local concentration of atmospheric carbon dioxide is decreased. Concerning that diffusion of gases from distant parts of the atmosphere might be relatively slow, one might assume that at a particular point the local concentration determines the availability of carbon dioxide for biomass growth. Therefore, Eq. (7.1) can be considered as a sufficiently good approximation for biomass production. Since the only part of biomass production that is dedicated to biofuels is taken into account, the progress of biofuel production can be considered proportional to the progress of dedicated biomass production. Obviously, the process of carbon fixation in the form of biomass is also proportional to the amount of grown biomass. It might also be assumed that the local concentration of atmospheric carbon dioxide is equal to the difference between CO_2 concentration before the start of biomass growth and that at any later instant of time. When the biomass-originated fuel is used, some fraction of carbon fixed in biomass is returned to the atmosphere causing an increase of CO_2 concentration. This recirculation may start at different time distances, t_z, from the start of biomass growth, also the rate of the recirculation process may be different than the rate of biomass growth (and consequently the rate of primary carbon fixation). The progress of recirculation is assumed to follow the same type of kinetic equation as the undisturbed biomass growth, i.e., it can be represented by

$$x_r(t) = \alpha a\{1 - \exp[-k_1(t - t_z)^2]\} \tag{7.5}$$

where $x_r(t)$ is the fraction of recirculated carbon at an instant of time, t, a—is the saturation level of the primary growth curve, α—is the fraction of produced fuel that at given instant of time is already used (burned). Figure 7.2 shows time dependencies of $x_r(t)$ as compared to unmodified, $x(t)$ for $a = 1$, $\alpha = 0.4$, the ratio $k/k_1 = 10$, and several values of $t_z/t_{1/2}$.

[1]The permanency is expected as a feature of the sustainable process.

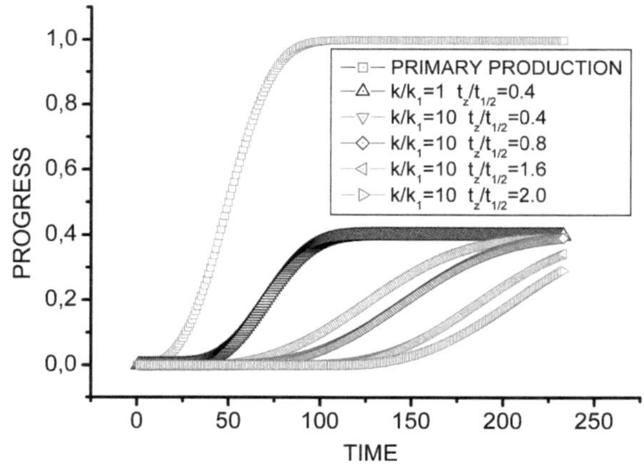

Fig. 7.2 The progress of the primary, and recirculation processes for several combinations of parameters

It is seen that the curves for recirculation processes reach their assumed saturation level $\alpha = 0.4$ at different instants of time dependent on both t_z, and k_1.

It can be further assumed that the amount of returned carbon dioxide may affect the rate of the primary process, because it increases its availability to the growing biomass. Therefore, the modified rate of the process may be expressed as

$$\frac{\mathrm{d}x_\mathrm{m}}{\mathrm{d}t} = a\{nkt^{n-1}\exp(-kt^n) + \alpha[1 - \exp[-k_1(t - t_\mathrm{z})^n]]\} \tag{7.6}$$

The examples of the time dependencies of the process rate which are affected by recirculation are given in Fig. 7.3.

The rate of the process occurring with recirculation, that starts at various instants of time, and also is slower than the primary one, during the initial period of time

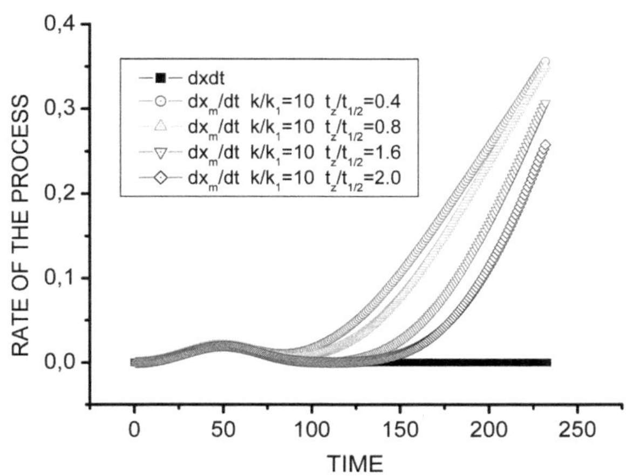

Fig. 7.3 The rate of the process occurring with superposition of delayed recirculation

follows the course of the rate of the unmodified process going through the local maximum at $t_{1/2,}$ and later rapidly increasing at times when recirculation becomes evident. Although it is not shown here, the upturn of the transition rate appears earlier for smaller ratios of k/k_1.

The progress of the process occurring with recirculation of some fraction of a substrate can be expressed in the form obtained by integration of formula (7.6):

$$x_{\mathrm{m}}(t) = a\int_0^t nk\tau^{n-1}\exp(-k\tau^n)\mathrm{d}\tau + \alpha\int_0^t \left[1-\exp\left[-k_1(\tau-t_z)^n\right]\right]\mathrm{d}\tau \quad (7.7)$$

Results of numerical integration of the above equation are shown in Fig. 7.4. The black curve shows the course of primary process (as in Fig. 7.2) showing leveling off at some instant of time, while the other curves correspond to the processes occurring when there is a return of some fraction of the substrate. In each case, the curve at some instant of time shows quite a steep increase in the progress of transition (production, growth, etc.). This upturn appears relatively early in the case $k = k_1$ (equal rates of the primary, and recirculation processes). In the other cases, this upturn appears later. Both characteristics play a similar role—when the recirculation process goes slower than primary one, $k_1 = 0.1k$, and when it starts later (an increase of t_z) the upturn is delayed. In the cases when recirculation starts relatively late the upturn is preceded by local leveling off corresponding to that occurring in primary transition curve. It can be concluded thereof that the faster is consumption of biofuel, the stronger capability of rebuilding the biomass growth potential.

It can be expected that in reality the growth rates of various types of biomass, like short rotation or normal forests, annual or perennial crops, as well as the rates of consumption of various biomass-derived fuels, are very different. Consequently, the real picture may be much more complex than this, presented above. Nevertheless, it

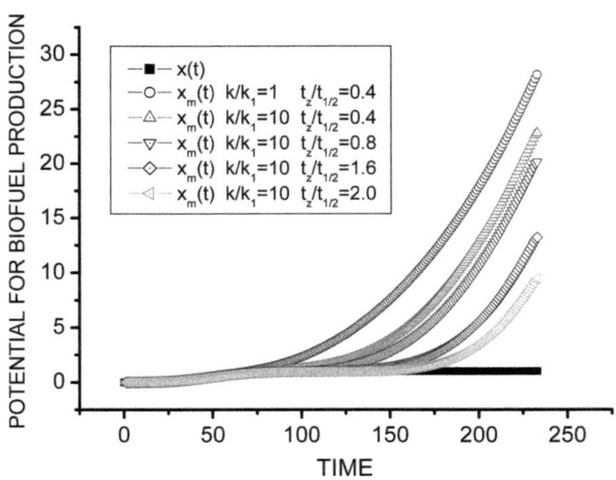

Fig. 7.4 Progress of the process with recirculation of a substrate

may be concluded that biofuel production does not spoil the natural carbon cycle, and does not violate sustainable development of the biomass production.

The second problem that concerns the sustainability of agriculture is whether or not the agriculture can be energetically self-sufficient, and to what extent it can contribute to the sustainable development of the whole economy. This can be analyzed on the basis of the results presented in earlier chapters. The question concerning energetic self-sufficiency of agriculture can be reduced to the question how big the fraction of arable land (or in a wider sense—substrate for biomass production) would have to be dedicated towards biofuels production in order to produce the amount of biofuel sufficient to cover demands of the whole agricultural production. Concerning values of, discussed earlier, energetic efficiency, ε, that at present state of technological and organizational development of agriculture, it would be difficult to reach the self-sufficiency without affecting the food production. It seems that using biomass wastes from agriculture itself, as well as the food industry, and decreasing fuel consumption in all production phases, i.e., agricultural operations, transport, industrial conversion, etc., the agriculture might reach situation closer to self-sufficiency, which would correspond to sustainability. The use of other alternative energy sources, like wind, solar, small water-powered electric generators, might be very helpful with this respect.

References

1. World Commission on Environment and Development Our Common Future (Bruntland Report). Oxford University Press (1987)
2. Alfredsson, E., Wijkman, A.: Prestudy The Inclusive Green Economy. Shaping Society to Serve Sustainability—Minor Adjustments or a Paradigm Shift? MISTRA, The Swedish Foundation for Strategic Environmental Research (2014)
3. Spreng, D.: Distribution of energy consumption and the 2000 W/capita target. Energy Policy **33**, 1905–1911 (2005)
4. Nigam, P.S., Singh, A.: Production of liquid biofuels from renewable resources. Prog. Energy Combust. Sci. **37**, 52–68 (2011)
5. Bharathiraja, B., Chakravarthy, M., RanjithKumar, R.: Biodiesel production using chemical and biological methods—A review of process, catalyst, acyl acceptor, source and process variables. Renew. Sustain. Energy Rev. **38**, 368–382 (2014)
6. Abnisa, F., Wan Daud, W.M.A., Husin, W.N.W., Sahu, J.N.: Utilization possibilities of palm shell as a source of biomass energy in Malaysia by producing bio-oil in pyrolysis process. Biomass Bioenergy **35**, 1863–1872
7. Russo, D., Dassisti, M., Lawlorb, V., Olabib, A.G.: State of the art of biofuels from pure plant oil. Renew. Sustain. Energy Rev. **16**, 4056–4070 (2012)
8. Siche, J.R., Agostinho, F., Ortega, E., Romeiro, A.: Sustainability of nations by indices: comparative study between environmental sustainability index, ecological footprint and the emergy performance. Ecol. Econ. (2007). https://doi.org/10.1016/j.ecolecon.2007.10.023
9. Odum, H.T.: Emergy in ecosystems. In: Polunin, N. (ed.) Environmental Monographs and Symposia, pp. 337–369. Wiley, New York (1986)
10. Beltratti, A., Chichilnisky, G., Heal, G.: Sustainable Growth and the Green Golden Rule. NBER Working Paper Series No 4430, National Bureau of Economic Research, Cambridge (1993)
11. Wasiak, A.: The effect of biofuel production on sustainability of agriculture. Biol. Syst. Open Access **5**, 171. https://doi.org/10.4172/2329-6577.1000171

Chapter 8
Epilog

During the last 20 years, the idea of sustainable development acquired official approval and political sense. One of the issues that is frequently discussed is the problem of global warming. This problem introduces contradictory opinions that appear cyclically. One side of the debate accuses human population, and especially the use of fossil fuels for the effects of global warming, whereas the other side claims no anthropological influence, and the only reason finds in the actions of Nature. It is not probable to achieve compromising solution in the short time. It seems, however, that if there is an even slight suspicion that some effects that might be dangerous to humanity are possibly caused by human activity—it is reasonable to mitigate the effects of that activity.

The production and use of biofuels is considered as a tool for mitigation of carbon dioxide emission, and, therefore, a tool for slowing down the thermal effects in the global atmosphere. Even, if the actual knowledge about global warming is not completely correct, the actions leading to the decrease of carbon dioxide emission are not dangerous to the atmosphere, and biological life. Those actions, however, might bring economic problems to countries having coal as a principal national resource. Consequently, all activities eliminating fossil fuels should be carried out with a tendency to optimize the path of realization.

On the other hand, if biofuel is produced to replace fossil fuels, and some amount of fossil fuels, at least temporarily, have to be used for biofuel production, the maximum care should be considered to assure that the amount of "fossil" energy used in biofuel production do not exceeds the amount of produced bioenergy. Otherwise, the whole effort does not make sense.

In several places of this book, it was shown that in some cases, the world's biofuels producers are close to this "limit of nonsense". It is, therefore, important to carry out research in order to identify the places of highest risk of loses, and to look for more effective ways of production. This is equally valid to agriculture as to transport, as well as industry converting biomass into biofuels. There is also a place for thinking about appropriate organization processes for reducing wastage of energy and resources in production processes.

A. Wasiak, *Modeling Energetic Efficiency of Biofuels Production*, Green Energy and Technology, https://doi.org/10.1007/978-3-319-98431-5_8

The present author strongly believes that computer simulations are a good tool to match the challenges of the present and future days.

There is still plenty of work to be done, and in many aspects, this book reveals gaps in the present knowledge and hopefully stimulates to further work.